SPC
SIMPLIFIED FOR SERVICES
Practical Tools for Continuous Quality Improvement

SPC SIMPLIFIED FOR SERVICES
Practical Tools for Continuous Quality Improvement

DAVIDA M. AMSDEN
HOWARD E. BUTLER
ROBERT T. AMSDEN

CHAPMAN & HALL
London · New York · Tokyo · Melbourne · Madras

UK	Chapman & Hall, 2–6 Boundary Row, London SE1 8HN
USA	Van Nostrand Reinhold, 115 5th Avenue, New York, NY 10003
JAPAN	Chapman & Hall Japan, Thomson Publishing Japan, Hirakawacho Nemoto Building, 7F, 1–7–11 Hirakawa-cho, Chiyoda-ku, Tokyo 102
AUSTRALIA	Chapman & Hall Australia, Thomas Nelson Australia, 102 Dodds Street, South Melbourne, Victoria 3205
INDIA	Chapman & Hall India, R. Seshadri, 32 Second Main Road, CIT East, Madras 600 035

First edition 1991

© 1991 Quality Resources

Originally published in the United States of America by Quality Resources, A Division of The Kraus Organization Limited

Printed in the United States of America

ISBN 0 412 44740 1

This edition not for sale in North America and Japan

All rights reserved. No part of this publication may be reproduced or transmitted, in any form or by any means, electronic, mechanical, photocopying, recording or otherwise, or stored in any retrieval system of any nature, without the written permission of the original publisher.
  The publisher makes no representation, express or implied, with regard to the accuracy of the information contained in this book and cannot accept any legal responsibility or liability for any errors or omissions that may be made.

On dedication page, scripture from the HOLY BIBLE, NEW INTERNATIONAL VERSION. Copyright © 1973, 1978, 1984 International Bible Society. Used by permission of Zondervan Bible Publishers.

British Library Cataloguing in Publication Data

Amsden, D.
SPC simplified for services: Practical tools for continuous quality improvement.
I. Title II. Butler, H. III. Amsden, R.
003

ISBN 0-412-44740-1

*This book is lovingly dedicated to
our grandchildren:
Weston and Vivian Miller
Julia and Alexander Butler
Nathaniel Amsden
Victoria Amsden
Edward Amsden*

*"I know the plans I have for you,"
says the Lord, ". . . plans to give
you a hope and a future. . . ."
—Jeremiah 29:11*

Contents

Acknowledgments ... xi

Introduction ... xiii
- The Need ... xiii
- What Is Quality? ... xiv
- How Do We Get Quality? ... xv
- Tools of Quality for Control and for Breakthrough ... xvii

MODULE 1
BASIC PRINCIPLES ... 1
- Basic Principles ... 2
- Causes of Variation ... 6
- Tools of Quality ... 7
- Variables Charts ... 10
- Attributes Charts ... 12
- State-of-the-Art Management ... 14
- A Criticality Designation System: Planning What Must Be Controlled ... 15
- Supplier Control: What's Coming in the Back Door? ... 17
- In-Process Control: What's Happening in Our Processes? ... 17
- Outgoing Quality Assurance System: What's Going Out the Front Door? ... 18
- Understanding Variability ... 19
- Decision Making and Problem Solving ... 19
- Employees and Problem Solving ... 21
- Management Uses of Quality Control Techniques ... 21
- Thought Questions ... 23

MODULE 2
BASIC PROBLEM-SOLVING TOOLS ... 25
- Brainstorming: A Downpour of Ideas ... 25
- What Is Needed for Brainstorming? ... 26
- How Does a Brainstorm Work? ... 27
- Prodding Techniques ... 29
- Completing the Brainstorm: A Thorough Soaking ... 31
- Difficulties with Brainstorming and What to Do About Them ... 32
- Cause and Effect Diagrams: Organizing the Brainstorm ... 32
- Why Use the Cause and Effect Diagram? ... 34
- How to Construct a Cause and Effect Diagram ... 34
- The Process of Constructing the Cause and Effect Diagram ... 37
- Types of Cause and Effect Diagrams ... 38
- Pareto Analysis ... 38
- How to Construct a Pareto Diagram ... 40
- How to Interpret the Pareto Diagram ... 46
- Summary ... 47
- Practice Problems ... 48

MODULE 3
QUALITY IMPROVEMENT TOOLS ... 51
- Flow Charts ... 51
- Process Flow-Chart Symbols ... 52
- Constructing a Process Flow Chart ... 53
- How to Use the Process Flow Chart ... 55
- Storyboarding ... 56
- What Is Needed for Storyboarding ... 57
- How Does Storyboarding Work? ... 59
- Evaluating the Storyboard ... 61
- Scatter Diagrams ... 62
- Where to Use a Scatter Diagram ... 63
- How to Construct a Scatter Diagram ... 63
- Summary ... 69
- Practice Problems ... 70

MODULE 4
FREQUENCY HISTOGRAMS AND CHECKSHEETS — 73
- What Is Variation? — 73
- Frequency Histograms — 74
- Constructing a Frequency Histogram — 75
- Some Cautions — 82
- What Frequency Histograms Tell You About Underlying Frequency Distributions — 85
- Frequency Histograms in Service Situations — 87
- Checksheets — 89
- Summary — 94
- Practice Problems — 94

MODULE 5
VARIABLES CONTROL CHARTS — 99
- Using Average and Range Charts That Are Already Set Up — 101
- Interpreting Average and Range Charts — 103
- Averages Outside Control Limits — 104
- Other Signs of a Process Out of Control — 105
- Sources of Assignable Causes — 106
- Ranges Outside Control Limits — 106
- Setting Up Average and Range Charts — 107
- How to Use Average and Range Control Charts in Continued Operations — 121
- Median and Range Charts — 122
- Developing a Median and Range Chart — 123
- How to Use Median and Range Control Charts in Continued Operations — 128
- Individual and Range Charts — 129
- Developing an Individual and Range Chart — 130
- Control Limits for Individual and Range Charts — 131
- Summary — 134
- Practice Problems — 136

MODULE 6
ATTRIBUTES CONTROL CHARTS — 137
- Why Use an Attributes Control Chart? — 138
- Percent Defective p-Charts — 138
- How to Use p-Charts — 138
- Interpreting Percent Defective p-Charts — 142
- Percent Defective, p, Inside Control Limits — 142
- Percent Defective, p, Outside Control Limits — 143
- Other Indications of Out-of-Control Processes — 145
- Types of Assignable Causes — 145
- Setting Up Percent Defective p-Charts — 146
- How to Use a Newly Developed p-Chart in Continued Operations — 157
- The np-Chart — 159
- Setting Up the np-Chart — 159
- c-Charts — 162
- How to Use c-Charts — 163
- Interpreting c-Charts — 164
- Setting Up c-Charts — 166
- Summary — 171
- Practice Problems — 171

MODULE 7
SAMPLING PLANS — 173
- Introduction — 173
- Acceptance Sampling — 174
- Sampling Risks — 177
- Applications of Sampling Plans — 177
- Sampling Plans — 178
- Parameters of Sampling Plans — 179
- Limitations of Sampling Plans — 180
- Sampling Techniques — 181
- Operating Characteristic Curves — 182
- How to Interpret an OC Curve — 184
- Developing an OC Curve — 185
- Acceptable Quality Level and Lot Tolerance Percent Defective — 190
- Changing the Acceptance Number — 191
- Changing the Sample Size — 193
- Average Outgoing Quality — 194
- Developing an AOQ Curve — 195
- Selecting a Sampling Plan — 198
- Summary — 203
- Practice Problems — 205

Contents ix

MODULE 8
SYSTEMS CAPABILITY 207
- Introduction 207
- What is Meant by Capability 208
- Defining the System 210
- Operation (Short-Term) Capability 210
- Average and Range Charts 210
- Limits for Individuals 212
- The Probability Plot 217
- Estimating the Proportion of Measurements Out of Specification 225
- System Capability 227
- Capability Index 228
- Summary 231
- Practice Problem 232

MODULE 9
PUTTING IT ALL TOGETHER 233
- Framework for Solving Problems 234
- Summary 237

Solutions to Practice Problems 239

Glossary of Terms 283

Recommended Readings and Resources 289

Appendix: Factors and Formulas 291

Index 293

Acknowledgments

SPC Simplified for Services grew out of the needs of our customers for a text like *SPC Simplified*, but one directed toward people in the service sector. This book, too, is written from a practical point of view, using real service examples shared by people who have the same kinds of problems you face on your job.

We wish to acknowledge the following: In Module 4, the discussion of the bucket of plastic-coated chips is based on the work of the late Dr. Walter Shewhart; the late Harold Dodge told the story of the munitions factory found in Module 5; the variables control charts used in Module 5 have been adapted for the service industry from forms distributed by the American Society for Quality Control. We developed the description of storyboarding from the work of Jerry McNellis and we thank Jack Nettles of the McNellis Company for permission to use this material. We also thank the Institute of Mathematical Statistics for permission to use material on the corner count test, which was developed by Paul S. Olmstead and John W. Tukey. Their article, entitled "A Corner Test for Association," was published in the Institute's journal, *Annals of Mathematical Statistics*, in 1947.

No one puts together a book like this without the contributions of a great many people. We would especially like to acknowledge the following people who helped us in several different ways. Some read portions of the manuscript and made many useful suggestions. Others shared their work experiences. So it is with pleasure that we give grateful thanks to Jonathan Amsden, Vivian Amsden, Tom Brown, Terry Comer, Gordon Constable, Jan Coon, James Derksen, Jackie Johnston, Richard Leavenworth, Jack Nettles, Linda and Jim Pressnell, Jack ReVelle, Linda Simons, Fran Wright, and the team at Quality Resources.

D.M.A.
H.E.B.
R.T.A.

Introduction

In this book, we talk about many of the tools and techniques of quality. These tools are easy to learn. So you can better see where the tools of quality fit in your service business, we will first talk about what it takes to become a quality organization.

THE NEED

In many manufacturing industries worldwide, quality is a major strategy for gaining the competitive edge. Quality in the Japanese auto industry, for example, means the quality of everything the company does as well as the quality of the product itself. This includes the quality of sales; of market research to find out what the customer wants, needs, and expects; of new product development; of ordering processes; of billing; of service of the product; etc.

What does this have to do with service organizations? Until recently, the really tough competition from abroad, especially from Japan, has been in manufacturing. Competition is now growing in service industries as well. Any company, whether foreign or domestic, that learns the ideas and tools of quality and practices quality management will become a serious competitive threat to your company.

Some examples may help you see the seriousness of your situation. A participant in one of our SPC seminars works for the Internal Revenue Service. We asked him why he was taking the seminar since his organization "obviously" has a monopoly on taxes. His reply was very enlightening. If the IRS fails to do a top-quality job, he said, Congress would consider hiring private companies to do the work instead of the IRS. Clearly, he did not think the IRS has it sewed up tight!

Another example is in banking. We find it very disturbing that the 10 largest banks in the world are foreign. Some of them have even opened branches in the United States.

A third example is from the University of Dayton (UD), where one of us teaches. Like many universities, UD is faced with declining enrollments. In order to attract and keep students, UD must compete with other universities. Part of its strategy is quality—to be a school of excellence.

Our point is twofold. First, no service industry, whether public or private, is safe from competition. Second, quality is often a major strategy used by your competition.

Quality plays two roles in service industries. First, quality is important to the customer. The customer wants quality in every aspect of a service. Was it delivered when expected? Was it at the agreed-upon price? Was the customer treated with courtesy and respect? Did the service do what the customer expected? Was the service free of errors in billing, pricing, quantity, etc.? All of these are important because they affect the customer's decision to do more business with your company. These aspects of quality also affect whether the customer will recommend or criticize your service.

The second role is one internal to your firm. The quality of your company's internal processes affects the costs of doing business, which in turn affect the ability of your firm to compete. Experts in quality point out that 15 to 30% of sales dollars in manufacturing are quality costs. That is, 15 to 30% of the sales dollars are used up by the costs of scrap, liability suits, rework, inspection, etc. But in service industries, quality costs run even higher! Here the costs are often 30 to 40% of the sales dollars. If you are in a service industry, your company's quality costs may be this high. Think of the savings that are possible from improved quality. If your competition pursues quality, they can beat the socks off you!

You may feel that things really aren't this serious. But think of how many manufacturing companies have closed plants or sent work "off shore." Think of the loss in market share suffered by the American auto industry. And then there's world banking. When a Japanese bank can increase its holdings in one year by the total size of the third largest bank in the United States, you can bet there's strong cause for concern!

There are companies that would like to eat you for lunch! Your competition won't wait for you to catch up. It's your job to compete, or get out of the business.

Quality allows a company to make great gains. By learning and applying the right tools and thinking, your company can improve. It can improve its ability to attract and keep customers, increase its market share, and be more profitable.

WHAT IS QUALITY?

Quality has numerous definitions. One way of looking at it is: Quality is whatever the customer defines it as. This may appear to be sidestepping the definition, but we don't think so. Bank One in Ohio administers thousands of surveys each year to find out what's important to the customer. Bank One does not say, "We know what quality is," but rather, "We must continue to ask the customer what quality is and how well we are meeting those expectations."

If we are to be competitive, our idea of quality must continually be defined by the customer. We must focus on the customer. This means we build and maintain our service organizations in a manner that assures continual focus on the customer. Recently, a speaker at a conference asked the audience: "Do you know who your new customers are this year? Do you know why they came to you?" Then he asked: "Do you know what customers you lost this year? Do you know why they left?" He then suggested that most companies could not answer; the audience's silence indicated he was right.

Who is "the customer"? Obviously, there are external or final customers who purchase our service. We can think of hospital patients, shoppers at a supermarket, patrons of a restaurant, or a vacationer renting a car.

We also have internal customers. The clerk who takes a handwritten or phoned-in order from the sales representative is a customer of the salesperson because he or she is the next person in the process. The clerk's need, want, and desire for clear, accurate information need attention from the salesperson, if for no other reason than to assure that orders are entered correctly. Simply because the job is done is not a guarantee that it will be done properly.

The flip side is that the sales representative is also the clerk's customer because the clerk is also providing a service. There is a two-way relationship between the sales representative and the order-entry clerk, and the quality of the work depends upon the quality of this two-way relationship.

To summarize what we've said so far, it's important to focus on the customer because it is the customer who defines quality. The customer is the final purchaser of your service, but remember internal customer relationships as well. We must strive to identify and focus on both internal and external customers.

HOW DO WE GET QUALITY?

Quality is a moving target. It's not something you can reach and then quit striving for. Quality is a moving target because your competition does not stand still. They get better and better. Furthermore, quality is, in a sense, perfection. We can continually strive for perfection, but on this earth, can we ever obtain it?

Seeing quality as a moving target, as perfection that is never attained, results in a process called continual improvement. *Continual improvement* is an operating philosophy driven by top management that says: "We will continually improve all of our services, all of our organization's processes. In short, we will continually improve everything we do." Once you understand the concept of continual improvement, everything in this book makes sense.

Continual improvement means we are constantly striving to improve everything about our organization. We do not begin by concentrating on the quality of our service, but rather on the customer to determine the customer's needs, wants, and desires. Then we design our processes to deliver services that meet these needs, wants, and desires. If our processes are good then our services will be good. We use the tools in this book to help maintain and improve our processes.

An example from retailing may help clarify what we mean. Management of a certain department store felt the store did not have the sales level it should. When management interviewed customers, they found out that the store wasn't offering what the customers really wanted. So the store changed its merchandise and management reorganized the layout of the store. The result was improved customer satisfaction and increased sales. In this example, the department store improved its process of listening to customers and using their feedback. Improving processes meant changing the merchandise and the store's layout. These improvements increased customer satisfaction and sales.

How can we accomplish continual improvement? A basic model of continual improvement is called *PDCA/PDCS*. PDCA stands for plan, do, check, and act; PDCS means plan, do, check, and standardize. Once you understand this basic model, you will see that each of the techniques we describe in this book helps us to do one of two things, either to control a process or to achieve a breakthrough or improvement.

The first step in PDCA is "plan" a project. The project could be a simple improvement in the software used to maintain life insurance policies. It could be a complex series of tasks involved in setting up a major new service. The second step is "do"; we carry out the plan. This could be a trial situation or an actual implementation where we produce the services and deliver them to the external customer. Next we go to the "check" stage. Here we determine how well we have accomplished the plan. At this point, we have a choice: Do we go on to the "act" stage or to the "standardize" stage? If we are satisfied with the results of our plan, we "standardize." In standardizing, we do what is required to keep our process going as is. For example, we might provide training for employees in our new software so that they can use it properly. These employees could use control charts to determine whether there are significant changes requiring correction in the use of the software. Control charts are among the tools we describe in this book.

On the other hand, when we check the implementation of our plan, we may decide that the "do" stage has not accomplished all we intended it to. Instead of going to "standardize," we must "act" or make an adjustment. This means we review our plan and its implementation and search for ways to improve. There are a variety of techniques we can use for improvement. Pareto analysis, brainstorming, cause and effect diagrams, and storyboarding—all of which are covered in this book—are four simple, yet powerful, techniques to use.

Now let's look at the PDCA/PDCS model a little differently. As in manufacturing, many companies in service industries have repetitive processes. That is, certain activities or services are performed over and over again. When we provide services, we are always doing one of two things: We are controlling our process or we are trying to make a breakthrough in our process. "Control" means preventing any change. For example, once an airline reservations agent is trained in how to handle a straightforward customer call, we want and expect that agent to continue processing calls as he or she was trained to do. We want the agent to "control" the process by keeping it the same.

Breakthrough, on the other hand, means deliberate action to achieve improvement. Suppose the airline's management finds growing customer dissatisfaction with how reservations are being handled. Management may decide to train reservations agents in improved ways of handling calls. This training could permanently improve customer satisfaction. Such a long-term improvement in customer satisfaction is an example of a breakthrough.

Once the breakthrough has occurred, we must immediately establish controls in order to maintain improved performance. Otherwise, we probably will not be able to keep the gains we've made. We said earlier that control and breakthrough represent another way of looking at the PDCA/PDCS model. Control is like the "check" and "standardize" stages in the PDCS cycle when we decide to standardize the improvement. Breakthrough is similar to the "check" and "act" stages in PDCA because we make deliberate changes—we act—based on the results of our check. We modify our plan accordingly; and we do or perform our service in an improved manner.

TOOLS OF QUALITY FOR CONTROL AND FOR BREAKTHROUGH

Some of the tools of quality are used for control; others are used for breakthrough. Some are useful for both. The problem-solving tools in Modules 2 and 3, for example brainstorming and flow charts, are excellent for breakthrough. In some cases, you can use the tools from these modules to determine why a process has gone out of control. We use frequency histograms (Module 4) for either control or breakthrough. Control charts, presented in Modules 5 and 6, are tools for control because they help us keep a process running without change. Sampling plans (Module 7) are best used where the process is not in what we call "statistical control." Quality is achieved by using some type of sampling plan to clean out errors or defects from the service after the service is performed rather than by preventing the errors in the first place. In such cases, sampling plans, or acceptance sampling, are a stopgap measure until management implements a process of continual improvement. Another use for sampling plans is for auditing the quality of your service. Notice that an audit is only telling what the current quality is; it is not controlling the quality. Systems capability (Module 8) gives us a clear idea of how well a process can do. Using this information, we can decide whether to control our process at this level of quality or whether we need to improve it by doing breakthrough activities. Module 9 ties it all together. Here we present a problem-solving framework that shows where the quality tools fit.

If you and your company learn and practice these tools and thinking, you will have the means to do something about your problems, to gain and satisfy customers, and to become a company of excellence.

MODULE 1

Basic Principles

NEW TERMS IN MODULE 1
(in order of appearance)

control charts	$\bar{X}$
frequency distribution	range
normal distribution curve	R
bell-shaped curve	upper control limit (UCL)
chance causes	lower control limit (LCL)
assignable causes	out of control
system causes	in control
inherent causes	percent defective chart
fishbone diagram	companywide quality control system
stable processes	
frequency histogram	criticality designation system
average	critical characteristics
process spread	critical operations
standard deviation	supplier control system
sigma	management by detection
control limits	in-process control system
variables chart	outgoing quality assurance system
attributes chart	management solvable problem
average and range chart	floor solvable problem

If you work in what is called the service industry, you may think your business is vastly different from the manufacturing industry. Manufacturers produce products that customers buy and use—products that may be useful for a short time or for many years. Your industry provides a service—which generally means you provide something for a customer and that's the end of it. The two industries may seem very different and there is some truth to this. The environment where services are provided for customers is usually very different from the environment where products are manufactured for customers. There are, however, similarities in the activities that must take place before a service is delivered to a customer or a product is produced for a customer.

Many years ago, when statistical quality control (SQC) was first being introduced in industries across the nation, it was common to hear manufacturing managers say, "I know they're using SQC in that business, but it won't work in ours." Today, many managers in the service industry are saying the same thing about statistical process control (SPC). Service industry managers need to learn what manufacturing managers learned many years ago. Any repetitive operation or process used to produce a product is subject to certain basic principles. These principles are the foundation for methods and procedures that can measure and monitor the performance of operations and processes.

Usually, services that are delivered to customers have predetermined standards—standards that have been determined by finding out what is acceptable to the customer. In order to make each service conform to its standards, services are produced through a series of repetitive operations. These operations have characteristics that are crucial to the successful production and delivery of the service. The quality of these characteristics determines the quality of the service produced. It matters not whether the service is cleaning rooms in a hotel, cashing checks at a bank, writing insurance policies, delivering food to a drive-through window, preparing and mailing monthly statements, or any of thousands of different services; the services are each produced through a repetitive process. These repetitive processes can be monitored and measured using statistical tools we call the "tools of quality." These tools, which are statistical techniques, can help us keep an eye on and control the quality of our process characteristics, and therefore the quality of our service. The basis for these statistical techniques is found in the six basic principles of statistical process control.

BASIC PRINCIPLES

The statistical techniques presented in this book use charts to present a picture of what is happening to and in a process as a service is produced. These are referred to as *control charts*. Many of these control charts are used to monitor the condition of the process. When properly used, control charts will tell you three things:

1. When you are doing something you shouldn't.
2. When you are not doing something you should.
3. When you are doing things right.

In short, control charts will indicate how consistent your job or process is. They'll show you:

1. When the job is running satisfactorily.

Or:

2. When something has gone wrong that needs correcting.

Control charts will provide you with "stop" and "go" signals. They'll enable you to "point with pride" or "view with alarm" (and look for the cause of the trouble!).

You must understand the basic principles underlying these statistical process control techniques if you are to use them efficiently and profitably. Many people feel uncomfortable when the word "statistics" is mentioned. It means using numbers and that means mathematics, but in using statistical process control techniques, you don't need to get deeply involved in mathematics. All you need to do is learn a few basic principles, and use some simple arithmetic and a calculator. An understanding of these principles will make it easier for you to understand and use the control chart techniques presented in this book.

All the ideas and techniques presented in this book are based on the following six principles. The first principle is:

1. No two things are exactly alike.

Experience has shown that things are never exactly alike. When two things seem to be alike, we often say they're "like two peas in a pod." But when we open a pea pod and take a close look at the peas, we see slight differences. The peas are different in size, shape, and color. If you are concerned with providing or delivering a service, you know that the service provided is never exactly the same each time either. In one way or another, the service will be slightly different in some element or part. If we measure those elements or parts of the service closely enough, we will see differences—no matter how hard we try to make them the same. No two things are exactly the same; they vary. We want to keep the variation between services as small as possible. To help ourselves do this, we use a second basic principle:

2. Variation in a product or process can be measured.

Some variation is normal in the delivery of a service. Over a period of time this variation tends to increase. If you make no effort to measure or monitor the variation normally expected in the processes used to provide your service, you could find yourself in a lot of trouble. That is, all processes that are not monitored "go downhill." Therefore, it is necessary to measure the output of any process to know when trouble is brewing.

4 SPC Simplified for Services: Practical Tools for Continuous Quality Improvement

When you check the output of a process or operation, you will quickly notice one feature. This feature provides a basis for the third basic principle:

3. Things vary according to a definite pattern.

If you want to see this pattern take shape, all you need to do is measure an element of the process you use to develop and deliver your service. If you record your measurements in tally form, you will see a pattern begin to form after you have recorded several of them.

An easy way to demonstrate this principle is to roll a pair of dice 50 or more times and record the total number of spots that come up on each throw. After a while, you will see a pattern begin to form. This pattern is sometimes called a *frequency distribution,* and is shown in Figure 1-1.

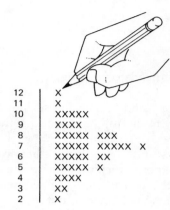

Figure 1-1. Frequency distribution.

A frequency distribution curve is formed by enclosing the tally marks in a curved line. This curve shows that there are more measurements or numbers in the middle and fewer as you go away from the middle. As you can see, the curve is shaped like a bell. See Figure 1-2. Not all kinds of measurements will give you a bell-shaped curve, but many will.

A frequency distribution curve will repeat itself whenever you take groups of measurements. This fact leads to the fourth basic principle:

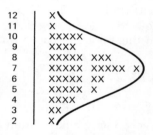

Figure 1-2. Bell-shaped curve.

4. Whenever things of the same kind are measured, a large group of the measurements will tend to cluster around the middle.

Most measurements will fall close to the middle. In fact, mathematicians can make a fairly accurate prediction of the percentage of measurements in various sections of the frequency distribution curve. This prediction is shown in Figure 1-3.

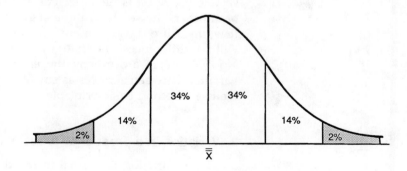

Figure 1-3. Approximate percentages of different measurements within the normal distribution curve.

How can we understand this curve in a practical way?

· If we measure certain aspects of the service you are offering to your customers, such as the time it takes to complete a transaction, the number of errors on forms or letters, or anything to which you can assign a number, and then make a tally of the measurements, we will eventually have a curve similar to the one in Figure 1-3.

- If we don't measure every item available, but merely grab a few and measure them, the chances are that 68 out of 100 measurements (34% + 34%) will fall within the two middle sections of the graph in Figure 1-3.
- The chances are that 28 out of 100 measurements (14% + 14%) will fall into the next two sections, one on each side of the middle sections.
- Finally, the chances are that 4 out of 100 measurements (2% + 2%) will fall into the two outside sections.

This may seem a little complicated, but don't worry about it. Just remember that measurements do tend to cluster about the middle, as shown by the bell shape of the curve. This particular curve is known as the *normal distribution curve.*

All this brings us to the fifth basic principle:

5. It's possible to determine the shape of the distribution curve for measurements obtained from any process.

By making a tally, or frequency distribution, of the measurements from a process, we can compare it with the range of measurements specified or required for that measurement (the "specification"). In this way we can learn what the process *is doing*, as compared to what we *want* it to do. If we don't like the comparison, we may have to change the process or the specification. The sixth basic principle helps us understand what must be done.

6. Variations due to assignable causes tend to distort the normal distribution curve.

A frequency distribution is a tally of measurements that shows the number of times each measurement is included in the tally. The tally you saw in Figure 1-1, which was made by rolling dice, is a frequency distribution.

A frequency distribution helps us determine when things are operating normally in a process. The *bell-shaped curve* fitted over the frequency distribution of dice rolls in Figure 1-2 is called a normal curve. If things are not normal, meaning something has gone wrong with the operation or process and should be corrected, the curve will be distorted and will not be a normal curve.

CAUSES OF VARIATION

The variation represented by a frequency distribution curve is the result of two types of causes. These are called by various names. In this book, we refer to them as *chance causes* and *assignable causes.*

Chance causes are those causes of variation that are built into the process. They cause the output of the operation to vary in a random manner. They are sometimes called *system causes* since they are present in the system or process at all times. They also are called *inherent causes* since they are inherent to the system. We normally consider chance causes as those causes of variation that the people doing the work can do nothing about. They are a part of the way the operation or process works. If we don't like the results of the operation and only chance causes are present, we must change the process to get different results. A change of this type is usually considered to be the responsibility of management.

Assignable causes are those causes of variation that distort the normal distribution curve. These are the causes that make the production of a service vary from the normal or bell-shaped curve in a nonrandom way. These causes of variation are not a part of the system. They occur from time to time and the source of these causes can be found and eliminated right at the point of occurrence. These are causes of poor quality service. As a person directly involved in the production and delivery of a service to the customer, you can do something about many of these causes. You, who are working at the point in the process where assignable causes happen, are the one who can improve quality by finding these causes and correcting them.

The source of variation in a process can be found in one or more of five areas. This is true whether the variation is due to chance causes or assignable causes. These five areas are as follows:

1. The *materials* used in the production and delivery of the service.
2. The *equipment* used to process and/or deliver the service.
3. The *methods* used in the production and/or delivery of the service.
4. The *employees** involved in the production and/or delivery of the service.
5. The *environment* in which the service is produced and/or delivered.

All variation in the service you perform for your customers is caused by variation in these five areas. This is illustrated in Figure 1-4.

The diagram in Figure 1-4 is sometimes called a *fishbone diagram*. It is very useful for searching out causes of trouble in a process. This will be discussed in detail in a later module.

If the variation in the service (caused by materials, equipment, meth-

---

*Note: when we use the word "employees" in this context, we mean *everyone* in an organization.

Figure 1-4. Fishbone diagram.

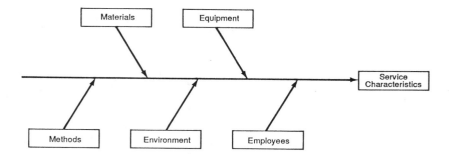

ods, employees, and environment) is due to chance causes alone, the service will vary in a normal, predictable manner. This process is said to be *stable*. It is important to know how the service varies under such normal, stable conditions—that is, when only chance causes are contributing to the variations. Then, if any unusual change occurs, we can see this change in our distribution curve. We can say it's the result of an assignable cause and not due to chance causes alone.

When assignable causes are present, the curve will be distorted and lose its normal bell shape. Figure 1-5 shows some typical distortions to the normal distribution curve.

Figure 1-5. Distortions of the normal curve.

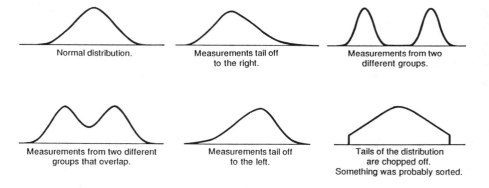

TOOLS OF QUALITY

To improve the quality of your services or to maintain the current quality, the processes that develop and deliver the services must be stable. The basic principles we have discussed are the foundation for the statistical techniques used to answer these questions:

- Are we doing something we shouldn't?
- Are we failing to do something we should?

- Is the process operating satisfactorily, or has something gone wrong that needs correction?

The techniques we will discuss in this book can be used on a one-time basis to solve specific quality problems, but great value comes from using them every day to control quality. They must become a way of life in our workplace.

Your goal should be to demonstrate that your job is operating in a stable manner. When the assignable causes have all been corrected, variation in the quality of your service is due only to chance causes.

When your operation is stable, then—and only then—can you know with confidence what your quality level is. When your operation is stable, your productivity and quality are as consistent as they can be with the process you are using.

Statistical methods of process control give you a way to picture and control quality using the "tools of quality." There are many tools of quality, but the ones discussed in this book are the easiest and most often used. These are the frequency histogram, the probability plot, and the control chart.

The *frequency histogram* is a graphic method of depicting what is happening in an operation at a given time. See Figure 1-6. The histogram is a simple bar chart that shows the frequency of occurrence of measurements. Using this chart, we can estimate the *average* of the process and the overall *process spread*. Although the histogram may show an operation is "not normal"—leading to the conclusion that an assignable cause is or was present—it does not have the ability to pinpoint the time when the assignable cause was acting on the process. Module 4 discusses the construction and uses of frequency histograms.

Figure 1-7 is the same curve shown in Figure 1-3, but with more information. The process spread represented by this frequency distribution curve is determined by the width of the curve. When the spread of the curve is wider than the specified or desired value of the service characteristic being measured, or when one tail of the curve goes outside of the desired value, we can estimate the number of times the service will fall outside the specification or fail to be of the desired quality.

The curve in Figure 1-7 also shows *standard deviations*. The sections of the curve are labeled here with the symbol σ *(sigma)*. This is the symbol for the standard deviation of the normal curve. The standard deviation is calculated mathematically and can be estimated in graphic form, but for now, let's just say it is a number that describes how the measurements cluster about the middle of the normal curve. In Figure 1-7, a normal curve is compared to the specification. The sections of the curve falling outside the specifications are shaded in Figure 1-7.

The histogram and the frequency distribution are used for such purposes as:

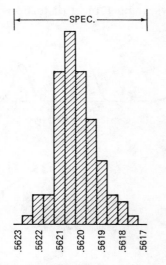

Figure 1-6. Histogram.

Figure 1-7. Frequency distribution.

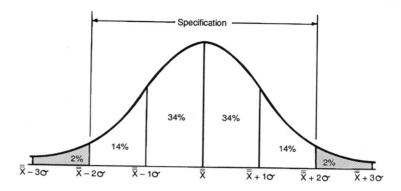

- Evaluating or checking processes.
- Indicating the need for taking corrective action.
- Measuring the effects of corrective actions.
- Comparing equipment or employee performances.
- Comparing materials.
- Comparing vendors.

The frequency histogram is a good statistical tool for the purposes just listed, but it is not a very good tool for monitoring the quality level of an ongoing process that produces a service. To monitor an ongoing process, we have another useful tool—the control chart.

A frequency distribution can be compared to a snapshot of a process and a control chart can be compared to a movie of that same process. A control chart is like a series of smaller pictures—a record of periodic, small inspections over time. Because the control chart is a running record of the job, it tells us when the job is running smoothly and when it needs attention. The control chart is a very good tool for showing you when you have a problem and when you have corrected a problem successfully. The control chart is an effective "tool of quality" because it uses *control limits*. Control limits are boundaries on a control chart within which we can operate. These limits are based on past performance; they show what we can expect from the process as long as nothing is changed. Each time the job is checked, we compare the results to the control limits. If the results are within the limits—no correction necessary. But if some of the points on a control chart fall outside the control limits, we know something has happened and the job is no longer operating normally. When this happens, we should take action to correct the situation.

In other words, control limits are warning signs that tell us:

1. When to take action.
2. When to leave the job alone. This doesn't necessarily mean you

don't have a problem. It just means you must take a different tack to identify the problem.

There are two general types of control charts.

1. *Variables Chart:*
   This type of chart is used where a characteristic is measured and the result is a number.
2. *Attributes Chart:*
   This type of chart is used where a characteristic is not measured in numbers, but is considered either "good" or "bad."

VARIABLES CHARTS

The *average and range chart* is the most commonly used of the variables charts. There are several kinds, and these are discussed in detail in Module 5. We will use the average and range chart for illustration.

The example shown in Figure 1-8 is a control chart of an operation that is evaluated by assigning values based on a demerit system. The service in this case is room cleaning in a hotel. Each task or element that must be completed to provide a satisfactory room-cleaning service has been assigned a value. If an individual task is not performed when cleaning the room, the value number of that task is recorded as a demerit. If the task is partially completed, a smaller demerit number is recorded. As an example, the task of cleaning the carpet could be assigned a value of 10. If the carpet has not been cleaned at all, the demerit value recorded would be 10. If the carpet has been partially cleaned, the demerit value recorded could be some number less than 10, depending on the degree of cleaning that was actually done. The total of all the demerit values for a room becomes the value for that room and is recorded on the control chart in the rows labelled "sample measurements." (In this example, five rooms were checked each day.)

This operation is just one of several types of services that could be evaluated, such as dry cleaning of clothing, carpet cleaning, or preparing insurance policies from agents' information. This control chart uses a five-piece sample, meaning five rooms were evaluated by an inspector each day. After the five values are recorded, they are added, and this value is recorded in the "sum" line. Next, the average of the five values is calculated by dividing the total, or sum, of the values by the number of rooms evaluated (five in this case), and entered in the "average" line. The average is plotted as one point on the chart. It is also called $\overline{X}$ (pronounced "X bar").

Another point to be plotted on the chart is the range. We find the *range* (or R, as it is called) by subtracting the smallest of the five values from the largest of the values. This number is recorded on the "range" line of the chart and plotted as one point on the range chart.

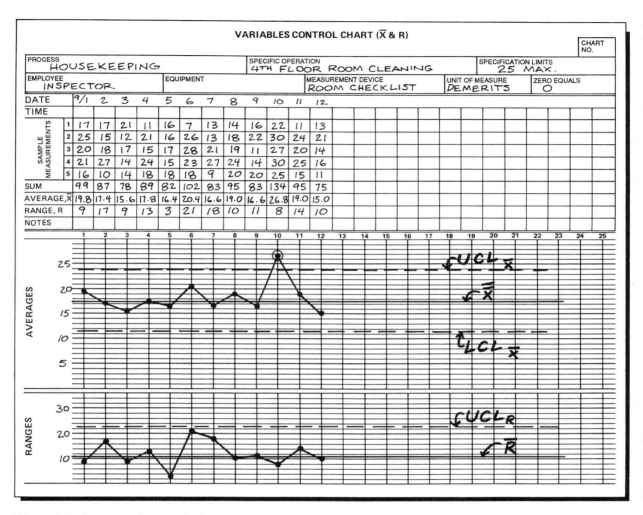

Figure 1-8. Average and range chart.

As you can see, the average and range chart is actually two charts. We use the average chart to monitor how well the operation is centered on the target value, and the range chart to monitor the spread of the values around the average.

On the chart you can see lines that are marked UCL and LCL. These letters stand for *upper control limit* and *lower control limit*. The upper control limit (UCL) and the lower control limit (LCL) are the limits within which we expect plotted points to stay. If a point falls outside one of these limits, the process is said to be *out of control*.

In Figure 1-8, we can see that an average point fell outside the upper control limit on the tenth day. This is a signal that something was distorting the normal distribution of the room-cleaning values. Some type of corrective action was probably taken because the process was back *in control* again on the eleventh day.

ATTRIBUTES CHARTS

We use attributes charts with "good/bad" inspection results. We do this by finding ways to assign numbers to these results. Remember, when we can assign a number to our information, we can use a statistical technique to monitor and control those numbers.

Several different attributes charts are available, and these are discussed in Module 6. Now, for a simple illustration, suppose the room-cleaning process in our average and range chart example was monitored so that the rooms were inspected merely to determine if they met a target value of 25 demerits maximum. Fifty rooms per day are inspected, and the results are recorded as a percentage of the 50 rooms not meeting the target. In other words, each room cleaned is either "bad" or "good."

An average and range chart cannot be used when only good/bad-type information is available. Instead, we use an attributes chart. One commonly used type of attributes chart is the *percent defective chart*, which is shown in simplified form in Figure 1-9.

The points plotted on this chart represent the results of inspecting 50 rooms after they were cleaned. The inspector evaluated each room and arrived at a demerit value. The number of rooms considered defective, or "bad," (those with a demerit rating over 25) was recorded on the control chart. The percent defective for each 50-room sample was calculated and plotted on the percent defective chart. Here, again, the points within the control limits indicate that everything is satisfactory, while the point outside the upper control limit (UCL) shows that something needs to be corrected.

Later in this book, you will learn how to develop these control charts and their control limits and also how to interpret them, but for now, just know that they are statistical tools based on a few easy-to-understand, basic principles.

Remember, when a process is in statistical control, the only variation seen in the process is due to chance causes. Even when services fall outside the desired standard but are still *inside* the control limits, the cause is still the result of chance. You will not find any assignable cause for this variation. The only way to solve such a problem effectively is to make a basic change in the process. That means doing something different from the way the process was originally set up, and that is usually the responsibility of management.

Control charts sometimes demonstrate that a process is operating in statistical control (that is, all points on the chart fall between the upper and lower control limits), but is still producing a service that does not

Figure 1-9. Percent defective chart.

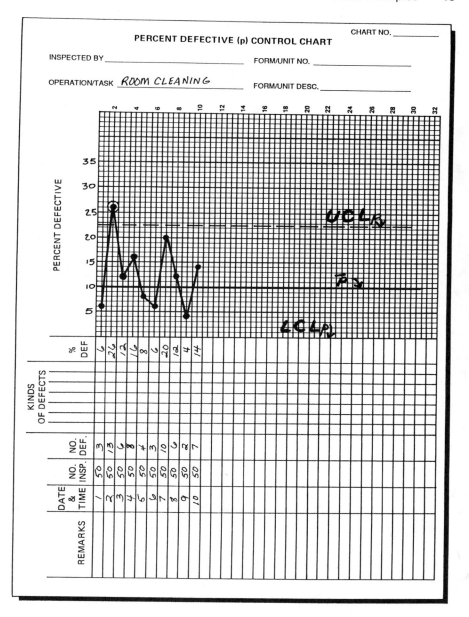

meet your desired standard. When this happens, it is a waste of time to try to find a solution to the problem by merely adjusting the process. (Adjusting the process here means putting it back to the way it was originally set up.) You should make an adjustment or take corrective action only when assignable causes are present.

On the other hand, if a control chart indicates that a process is out of statistical control, it means that an assignable cause is present that can be found and corrected. This is the type of problem you face every day on the job.

If you intend to use control charts to help maintain and improve the quality of your work and the services you produce, you must learn to think statistically. All the statistical techniques you will encounter in this book have their foundations in the basic principles we have been discussing. An understanding of these principles will help you to understand and use statistical process control charts.

Those who are uninformed in the use of statistical process control may say, "This is all well and good, but can I use it in *my* work area?" The answer can be given in one word: Yes! You can use a control chart on anything to which you can assign a number.

STATE-OF-THE-ART MANAGEMENT

The following modules of this book tell you how to use the tools of quality, including control charts and problem-solving techniques. But first, it is important to understand what it takes to manage a successful quality assurance system, in combination with those tools of quality.

This section of Module 1 provides an overview of quality methods and discusses how a company could apply these methods to an overall quality control system. Most of this book explains how to deal with problems of quality in the delivery of service. In this section, we touch on how all these quality activities relate to the activities and responsibilities of those who supervise and manage the company.

Even though a company can be quite successful in the marketplace, success and profits cannot be maintained for long without good managerial practices. Many companies go out of business because the competition provides a better quality product or service. Studies show that many companies that fail have as much technical know-how as those that succeed, but are simply beaten in productivity and quality, and therefore in price. This means that even though "failed" companies had the same technical knowledge as their competitors, they didn't manage their businesses as well as their competitors.

Close examination has shown that the management philosophy of employee involvement in quality control is a key factor that enables successful companies to outproduce their competition and still maintain better quality.

State-of-the-art managing—a structured approach to continuous improvement—is quickly becoming a way of life in the large manufacturing companies in this country and is just beginning to be seen in service companies. Managers are beginning to realize that if their businesses are to survive, this philosophy must be installed at all stages of the process of producing and selling services, from the smallest supplier to the customer.

State-of-the-art managing calls for suppliers to provide the best materials and services. Many large companies simply will not buy from suppliers who do not maintain high quality. It's likely that many of your customers are establishing high standards for the services they use, which will force companies like yours to concentrate on providing high-quality output. Some of these standards are based on the techniques discussed in this book. They include:

- top-management commitment to quality, including a timed action plan with supportive resources
- management action to provide systems and environments that promote employee involvement in continual improvement of quality and productivity
- quality training for management and all other employees, including inspectors
- use of statistical methods to verify that the processes of developing and delivering services are stable, capable, and on target with requirements
- use of statistical process control to meet customer requirements, including charts to measure process capability and ongoing process control
- a system for assuring that purchased material/service meets requirements. This system includes the use of statistical control charts.

A good quality control system has many elements. In fact, we can say that all the activities of a company are elements of its quality control system because the people who perform the activities needed to produce a service are (or should be) concerned with the quality of the service. When everyone cares about and is involved in quality, the company truly has a *companywide quality control system.*

An effective quality control system includes four elements that work together:

criticality designation

supplier control

in-process control

outgoing quality assurance

A CRITICALITY DESIGNATION SYSTEM: PLANNING WHAT MUST BE CONTROLLED

Before a new service or product is offered to potential customers, you must plan. You must plan for the equipment, the materials and supplies, and how you will verify the quality of the service. Materials used in providing a service and the elements or characteristics of the service are usually specified as a part of the service. It is not always clear, however, what should be monitored during the development and delivery of the service. People often ask, "What do I check?"

An easy answer is that everything must be checked. To be practical about running your business, however, you know this is not economically feasible or possible. As a producer of services, you need a system to determine which characteristics or elements of your service are most important to your customer. This is called a *criticality designation system*. It identifies those elements that require the use of statistical control techniques during the production and delivery of services.

The primary requirement of any service, from the simple delivery of a sandwich to the complex production and delivery of financial services, is that it must meet the customer's needs and expectations. The service must do what it is supposed to do. If you produce and deliver the service to an end user or customer, you may be able to evaluate it before it's delivered; but if you operate only a part of the process and supply only some elements of the service, it's not always possible to evaluate how well you are meeting the customer's needs and expectations. You must decide what to check among the various activities performed in your area of the business.

When faced with this decision, you can do one of several things. You can check every characteristic of every element of the service you produce (a good way to go out of business), or you could choose not to check anything and let the customer do the checking and evaluating of the quality of your service (another good way to go out of business). Obviously, the best approach lies somewhere between these two.

To make the best possible decision, you must know the function or functions of the elements of the service you are producing. What is your service or part of a service intended to do? If you know what your product or service is intended to do, it should not be difficult to determine which characteristics or elements are most important in enabling the service to do its job. The characteristics of any service that enable that service to perform its designated function are called *critical characteristics*. If a critical characteristic does not meet its specified design requirements, the service will not perform its designated function as it was intended. The operation that produces such a characteristic must be controlled to assure that all characteristics from that operation do indeed meet the specified design. The best way to do this is to use statistical techniques in the production of these characteristics.

Whether your service is simple or complex, or whether your organization is large or small, you will find it necessary to identify the critical characteristics or elements of your services. This is part of planning what must be controlled.

Critical operations are those operations that create critical characteristics. These operations must be controlled closely and maintained statistically stable to assure that the characteristics from those operations are stable and within specification.

SUPPLIER CONTROL: WHAT'S COMING IN THE BACK DOOR?

Most service companies find it necessary to purchase the materials or services they use in the production of their services. Many companies receive materials such as purchase orders, claim forms, field requests, etc., from other parts of the organization or from suppliers outside the company. These companies should know the quality of the materials they're receiving. Do they meet the needs of the process of developing and delivering your services? In other words, what is the quality of the materials and information you receive to produce your services?

Your customers expect you to provide services that have been produced in a state of statistical control. To be successful in doing this, you must know, in turn, that the materials and data provided to you by *your* suppliers have also been produced in a state of statistical control and that they conform to your specifications.

A more effective way to monitor the quality of materials coming into your company is to have a *supplier control system* that uses the criticality designation system. If you have identified the characteristics critical to the function of your services, pass this information on to your supplier. The material you receive from other locations or suppliers is essential to the quality of the services you provide; therefore, your suppliers should be required to use statistical techniques to monitor and control the critical characteristics of the materials supplied to your company.

A supplier control system should assure the quality of the materials you purchase to use in the production of your services. You can accomplish this most effectively with a system that requires your suppliers to show you that their critical operations are statistically stable and capable of producing material to your specification. Your suppliers should maintain control-chart evidence of the performance of critical operations. You can review these charts from time to time to be assured of continued good performance by suppliers. The control charts will also show when timely action was taken to prevent the shipment of nonconforming material. When a supplier control system is in place, the inspection of incoming materials can be held to a minimum. Then, and only then, it is reasonable to use incoming materials in your service system without any inspection.

IN-PROCESS CONTROL: WHAT'S HAPPENING IN OUR PROCESSES?

This book describes practical techniques for measuring, monitoring, evaluating, troubleshooting, and improving the processes that produce your services. These are the essence of in-process control.

Many companies use systems of inspecting or testing to maintain (what they believe is) control of the various processes or operations used to develop and deliver their services to their customers. Many supervisors will state that everything in their department is "in control." By this they really mean that as far as they know, there are no problems. In statistical process control terminology, the phrase "in control" means something different from the traditional definition. When the term "in control" is used in a company with statistical process control, it means there are no assignable causes present in the processes. It does not necessarily mean that the services being delivered to the customer are what the customer wants or needs; it just means the processes are producing the services as they have been designed to do.

Any system based on merely inspecting or checking is sometimes called *management by detection*. Supervisors or managers using this system of management spend their time trying to determine if the services being delivered meet the specifications. They have no way of telling whether the nonconforming services they find are due to the inherent variation in the process or to an assignable cause. They attempt to assign a cause to each nonconformance, when, in fact, about 85% of nonconformance is actually due to chance causes built into the process. Only about 15% of nonconformance is due to assignable causes.

State-of-the-art managers know that the best way to avoid spending time looking for assignable causes that don't exist is to *prevent* the creation of services that do not conform to the specifications of the customer. You will learn in the following modules that statistical process control techniques can be used to predict when a process is headed for trouble. These techniques tell supervisors when to take corrective action and when to leave a process alone. An *in-process control system* that assures corrective action when it is needed is very effective in preventing the creation of nonconforming services. It works far better than a system that merely detects some poor services.

The successful in-process control system must measure process capability and detect when that capability has changed. The statistical techniques described in this book are normally used for these purposes. The state-of-the-art manager uses these techniques to develop an in-process control system based on the criticality designation system. The criticality designation system identifies the characteristics of a service that must be controlled, and the in-process control system uses statistical process control charts to monitor and control the operations that create those characteristics.

OUTGOING QUALITY ASSURANCE SYSTEM: WHAT'S GOING OUT THE FRONT DOOR?

Any organization providing services to customers needs some additional assurance that the services being provided to their customers on an ongoing basis conform with the customers' needs and expectations. To provide this assurance, many companies use quality assurance auditors. These auditors are not inspectors who merely look for deficiencies in the service. The function of these auditors is to make certain the supplier control system and the in-process control system are working as they were designed to work.

Like the in-process control system, the *outgoing quality assurance system* uses statistical tools to determine whether processes are running satisfactorily or whether something has gone wrong and corrective action is needed. Data gathering at this phase of the process often reveals trends, and the manager who acts on these trends can head off quality problems over the horizon. Attributes control charts can be used to determine the quality and, more importantly, the stability of the processes. Frequency histograms and probability plots can be used to compare variable characteristics to the specification and to estimate outgoing capability. Outgoing quality assurance serves as a double check to make certain that all the other parts of the quality control system are working efficiently.

If the system breaks down and quality worsens, the quality assurance auditor may be your last chance to detect trouble before it gets to your customer. Quality assurance auditors use the tools of quality described in this book to audit and verify the continuing capability of the processes that produce and deliver services that meet the wants, needs, and expectations of the customer.

UNDERSTANDING VARIABILITY

To successfully manage any part of an organization that provides a service in the marketplace, you must first understand the variability that exists in those services. Remember, the first basic principle of quality control is *no two things are exactly alike*.

The purposes of the statistical techniques discussed in this book are to measure and to evaluate the variability of the characteristics of the services you produce. The mathematics of statistics is quite complex, but you do not need to become deeply involved in the mathematical basis of statistics. You must, however, understand the basic concepts of statistical analysis. If you want a greater understanding of statistics, other books on this subject are listed in the back of this book.

DECISION MAKING AND PROBLEM SOLVING

Some decisions that affect quality are made long before production of the service starts. These decisions involve the design of the process, the equipment, the materials, and the methods to be used. Along with the employees who work in the systems, these elements determine the inherent variability of the process. This is "normal" variation, which is built into the process and is responsible for chance causes of variation.

Employees are not usually involved in decisions made before production starts. For this reason, poor quality resulting from the inherent variation of a process is usually referred to as a management responsibility or a *management solvable problem*. Employees can only maintain a process at the level of variability built into the process by management. If inherent variability is to be reduced (thereby improving quality), management must reduce it.

Variations not inherent to the process come from changes in the materials, methods, employees, and equipment that were originally set by management. This kind of variation distorts the normal pattern of variation. It is not due to chance causes built into the process. The causes of this variation, called assignable causes, can be identified and corrected. Problems due to assignable causes are called *floor solvable problems* because they come from something other than the inherent variability of the process and they can be corrected "on the floor" at the point at which they are detected.

Good managers maintain decision-making authority at the lowest possible level, and employees can and should identify and solve quality problems that arise from assignable causes. If employees are given this responsibility, they must have the tools for making good decisions.

Statistical process control (SPC) techniques are highly effective. The purposes of statistical process control are, first, to identify and eliminate floor solvable problems and then, when processes are stable, to attack the management solvable problems.

When an assignable cause is determined to be present, corrective action is required to identify the cause and eliminate it. This must be done as soon as possible for two reasons. First, assignable causes are not a part of the system of chance causes built into the process. They do not operate on a continuing basis, but rather come and go. If corrective action is not taken in time, an assignable cause may come and go many times before it is corrected and the overall quality of the service may suffer. Second, when an assignable cause is present in a system, the estimates of the variation in the output of the process are no longer valid. The capability of the process to produce the service within specification is no longer known. (See Module 8 for a discussion of capability.)

The SPC techniques discussed in this book are the tools that employees need in order to make the decisions about quality that are rightfully their responsibility. They can use these techniques to determine when everything is normal and a process can be left alone, or when something is wrong and needs corrective action. Managers need the skills that will enable them to review the various SPC charts being used and to determine that the proper decisions are being made. (See Modules 5 and 6 for discussions of these control charts.)

EMPLOYEES AND PROBLEM SOLVING

A state-of-the-art manager makes certain that employees are trained to find and remove assignable causes through control chart techniques and to solve problems using problem-solving tools. In addition, such managers listen to employee suggestions for ways to reduce the inherent variation in the process. With training and encouragement, employees are able to make valuable contributions to correcting floor solvable problems as well as management solvable problems. State-of-the-art managers themselves use problem-solving techniques, such as Pareto analysis, brainstorming, flow charts, storyboarding, and cause and effect diagrams, to attack the management solvable problems.

MANAGEMENT USES OF QUALITY CONTROL TECHNIQUES

Control charts are not just for the use of employees. Managers have many uses for them. Managers may not often acquire data and plot them on control charts. In fact, they may seldom analyze charts for the purpose of taking corrective action. Such actions should be the responsibility of employees. Even so, managers will find it very valuable to have a thorough understanding of all types of control charts.

Control charts identify the part of the overall quality problem that's due to assignable causes. In doing this, they also identify the part of the problem that's due to the inherent variability of the process—the management solvable part.

The assignable causes and floor solvable problems must be identified continuously so that managers can successfully attack the management solvable problems. An efficient way to do this is to establish a regular procedure in which managers review the control charts that are generated during the production of a service. Such a procedure is a good management tool because it accomplishes several things at once:

- A review of completed control charts indicates how large (or small) a quality problem you have due to assignable causes. These causes result in poorer quality and lower productivity than expected. When a control chart shows a process to be statistically stable—that is, when there are no assignable causes present—it can be said with confidence that the process is generating its most consistent quality at maximum productivity.
- A system for management review of completed control charts demonstrates a commitment by management to quality. When such a system is in effect, employees know that management is truly interested in their efforts to improve quality.
- Reviewing control charts will help management determine whether incoming data or materials need to be improved. Properly com-

pleted control charts contain notations that tell when problems with incoming data or materials caused points to fall outside the control limits. As such, they can serve as an effective part of a supplier control system.
- As for in-process control, managers will be able to judge more accurately how well employees and supervisors are maintaining equipment. Control charts can tell whether preventive maintenance was performed before services were produced, or whether equipment breakdowns prevented employees from meeting schedules.
- When monitored over a period of time, completed control charts also show whether assignable causes have been corrected or whether the same causes continue to occur.
- Control chart review also allows management to verify the correctness of their decisions. If management has decided to change to a different supplier of materials, the control charts will show whether or not the service has improved. If a piece of processing equipment is replaced, the control charts will show whether the decision to replace the equipment was sound.
- If production of services must be reduced and a number of pieces of equipment are used to perform the same operation, a review of the control charts will make it easier for management to decide which equipment to shut down.
- If customers complain of problems relating to the quality of a service, control charts generated during the time that service was produced can help management assess the validity of the customers' claims.

All these benefits—and more—are gained when management uses control charts as tools for making better decisions. Opportunities for applying statistical control charts should not be overlooked. Control charts are not intended to be used solely by employees for maintaining quality and solving quality problems when they arise in the production of services. Managers should not feel they have no need to develop any skills in understanding and using control charts. Nothing could be further from the truth—control charts are equally useful to managers and to those who produce the services.

Companies using the prevention philosophy of management—based on the sound statistical practices discussed in this book—will be able to hold their own in a changing economy and to compete successfully in the world marketplace.

THOUGHT QUESTIONS

1. How closely does your company follow the requirements for state-of-the-art management?
2. Does your company have a criticality designation system? If so, what benefits do you see?
3. How would a supplier control system benefit your company?
4. Does your company have an in-process control system? How effective is it?
5. Does your company have an outgoing quality assurance system? How effective is it?
6. Does management in your company understand the concept of variability?
7. How would an understanding of variability help you in achieving and maintaining high quality in your organization?
8. Does management in your company understand its responsibility for dealing with the inherent variability in its processes?
9. What is the usual practice in your company for dealing with poor quality resulting from the inherent variability in a process?
10. Does management in your company see the importance of understanding and using control charts? Do you?

MODULE 2
Basic Problem-Solving Tools

> **NEW TERMS IN MODULE 2**
> (in order of appearance)
>
> chronic problem
> sporadic problem
> brainstorming
> piggybacking
> cause and effect (C and E) diagram
>
> process cause and effect diagram
> Pareto analysis
> Pareto diagram
> cumulative frequency
> cumulative percentage

Problems crop up in almost every situation. Even though your job goes smoothly most of the time, haven't you asked yourself, "Why have there been so many customer complaints this month?" "Why is housekeeping always short of linens on the weekend?" If you ask questions like these, you have opportunities to solve problems and to make changes.

In this module, you will learn three methods that will help you solve your job-related problems. These methods are useful for solving *chronic problems*, where the same thing happens over and over again. For example, you may find that monthly reports are typically many days late or there are often coding errors on bank statements. (Another type of problem, the *sporadic problem*, happens only once in a while, as when there is a power failure or a dishwasher breaks down. In this module and in the next, however, we are concentrating on ways to solve chronic problems.)

Finding solutions to chronic problems, the ones that keep happening, will result in breakthroughs to new and better operating levels and so will bring about improvements in quality, productivity, and service. Success in solving problems brings real satisfaction.

BRAINSTORMING: A DOWNPOUR OF IDEAS

In the problem-solving process, you need to identify problems as well as determine their causes. *Brainstorming* helps to do both. It's an excellent way of identifying problems, such as the ones you see on your job, and it's also a good way to gather many possible explanations for a specific problem.

Brainstorming is a group problem-solving method. It taps people's creative ability to identify and solve problems because it brings out a lot of ideas in a very short time. Because it is a group process, it helps build people. For example, brainstorming encourages individual members to contribute to the working of the group and to develop trust for the other members.

WHAT IS NEEDED FOR BRAINSTORMING?

A group willing to work together.

In order to begin a brainstorm, you need a group of people who are willing to work together. This may seem impossible, that the people you work with can never be a team. However, brainstorming can be a key to building a team! Furthermore it is a great tool for the group that is already working together.

Who should be included in the group? Everyone who is concerned with the problem, for two reasons. First, the ideas of everyone concerned with the problem will be available for the brainstorm. Second, if those people take an active part in solving the problem, they will be more likely to support the solution.

A leader.

Anyone can lead a brainstorm: the boss, one of the regular members, or even an outsider. The important thing is that there *must* be someone who can and will lead.

The leader is needed to provide guidance so that the brainstorm produces ideas. The leader should exercise enough control to keep the group on track while encouraging people's ideas and participation. He or she lays aside personal goals for the benefit of the group. In this sense, the leader both leads and serves, and walks a fine line between participation and control.

A meeting place.

The group needs a place where they will not be distracted or interrupted. In some facilities, there are rooms set aside for group meetings. In other situations, groups use a supervisor's office, a corner of the break room, or even an executive conference room.

Equipment.

The group needs flipcharts, water-based markers, and masking tape for posting the charts on the wall.

HOW DOES A BRAINSTORM WORK?

Following these general rules can facilitate a productive brainstorming session.

1. Choose the subject for the brainstorm.
2. Be sure that everyone understands what the problem or topic is.
3. Each person takes a turn and contributes one idea. If somebody can't think of anything, he or she says "Pass." If someone thinks of an idea when it is not his turn, he should write it down on a slip of paper to use at his next turn.
4. Have a recorder who writes down every idea. Be sure the contributor of the idea is satisfied with the way it is written down. The recorder should also have an opportunity to give his or her idea.
5. Write down *all* the ideas.
6. Encourage wild ideas. They may trigger someone else's thinking.
7. Hold criticism until after the session—criticism may block the free flow of ideas. The goal of brainstorming is quantity and creativity.
8. Some laughter is fun and healthy, but don't overdo it. It's O.K. to laugh with someone, never at them.
9. Allow a few hours or a few days for further thought. The first brainstorm on a topic will start people thinking. An incubation period allows the mind to release more creative ideas and insights.

To begin a brainstorm, review the rules and start. You can place a time limit on the brainstorm, perhaps from 12 to 15 minutes. This may seem too short a time, but it is better to begin with short periods and lengthen them as your group becomes used to working together. Then you can use longer sessions productively when they are needed.

Let's illustrate brainstorming with a skit. This group includes five people: Lynn, the leader; Brady; Laura; Nate, the recorder; and Ed. Since they have been meeting for only a short time and the members have not had much experience with brainstorming, the leader has to do most of the work of keeping them on track. As the group gains experience, other members should begin to share leadership responsibilities.

Lynn: Last time our group decided to work on the causes of so many billing errors. Let's brainstorm the causes of this problem. Nate, you're good at the flipchart. Would you be recorder this session?

Nate: O.K., as long as you don't complain about my writing. (Laughter.)

Lynn: You write as well as I do! Besides, you can spell. Our topic, which Nate is writing down, is "Too many billing errors." Let's put a time limit of 15 minutes on this session. And don't forget the rules:
- We go around from person to person, one idea at a time.
- Don't worry if your ideas seem off the wall. Even if an idea is a wild one, it could trigger someone else's thinking.
- No evaluations. No "That's great" and no "That's dumb." There will be plenty of time later to look at the ideas.

O.K. Are we ready? (All agree.)
Brady, you start.

Brady: Computer program for billing. (Nate writes "Billing program.")

Lynn: Better add the word "computer" to make it clear. (Nate adds the word "Computer.")

Laura: I've noticed that a lot of our trouble is coming from customers in region 2. And I think there is a problem with the trunk lines. . . .

Lynn: Remember, we give one idea at a time. Jot down the other idea so that you don't lose it. Let's say, "Customers in region 2." Is that O.K.? (Laura nods.)

Lynn: Nate, it's your turn.

Nate: I think I'll pass this time.

Ed: Lack of training.

Nate: Now wait a minute. I was responsible for the training on the new billing system and are you telling me I. . . .

Lynn: O.K. guys. Let's hold discussion and evaluations until later. We need to bring out as many ideas as we can. It's my turn. I'll build on the idea of "customers in region 2." Maybe it's only one or two of them. Nate, write "One customer."

Brady: Pass.

Laura: Trunk lines. Customers are having trouble getting through to systems service. They are put on hold or are disconnected. By the time I talk to them, are they mad!

Nate: How do I write that? "Trunk lines inadequate" or what?

Laura: "Trunk lines inadequate" catches my idea.

Ed: The new billing program has so many bugs in it we should call an exterminator. (Laughter.)

Lynn: Let's think about that. Ed may have something there. Nate, write "Bugs in program."

And so it goes. The brainstorm is fun and creative. Everyone is involved. Evaluations and discussion are postponed. Healthy laughter is accepted. As the team develops its list of possible reasons for the problem, the brainstorming process will give valuable guidance in finding one or more major causes. This group has begun to work together in solving the problem.

PRODDING TECHNIQUES

Sooner or later the downpour of ideas in the brainstorm dries up. What do you do to get it going again? Or what do you do with a silent member who doesn't participate? In the following sections, we will suggest some ways to deal with these problems. But be careful! Know your group. And remember that you are developing a group process, which is exciting and powerful, but fragile.

Encouraging Ideas: Priming the Pump Again.

If a brainstorm seems to be slowing down, the leader may suggest *piggybacking*. Piggybacking is building on others' ideas. In the skit, Lynn suggested "one customer" as a piggyback to Laura's idea of "customers in region 2." To give another example of piggybacking, one kind of "bug in the program" might be a "decimal place that isn't entered" when the user expects it. Ed's joking idea could lead to a useful one.

Another technique is to suggest opposites. The leader says, "Can we expand the number of ideas on the flipchart by giving opposites?" If somebody says "billing too much," the opposite would be "not billing enough."

You can also try quick associations. The leader gives a word or phrase. The members respond as quickly as possible with an associated word that could apply to the problem in some way. For example:

Lynn: Order.

Nate: Form.

Laura: Complicated.

Ed: Backlog.

Brady: Swamped.

Lynn: Now try a new phrase. Data input.

Nate: Messy form.

Laura: Hard to read.

Ed: Pass.

Brady: Repeats.

Lynn: Now let's get back to the main brainstorm. Brady?

Brady: Sure! Maybe we have billing errors because the data entry form is too complicated to fill out easily when we take information over the phone.

The leader may also prod the brainstorm by tossing out ideas in certain directions. He or she may ask the group to work on a new area of possible causes or pursue one area in greater depth. We might see something like this:

Lynn: So far we've looked at a number of general topics. Now let's consider the training issue that was mentioned earlier.

Ed: Hey, Nate. You did a super job when you led the training sessions. That guy from the head office was the problem. He never explained the new system in a way *I* could understand it. So write "Trainer."

The Silent Member.

When a member of the group doesn't speak up, the best advice we can give is "Be patient!" Sometimes a member will be very quiet for meeting after meeting, and then he or she will begin to open up. When this happens it's very exciting, so give that person time. Maybe he or she will always be quiet, but will serve the group in other valuable ways. Once we invited a brainstorming group to give a presentation. The quiet member had the job of operating the overhead projector. Later, he was the one who privately gave us a compliment. We suspect he does the same kind of building-up with his group.

There's a simple but effective method to help bring out the silent member. Remind the whole group that when each person's turn comes in the brainstorm, he or she just says "Pass" if not ready with an idea. That gets people off the hook, but it also breaks the sound barrier: they hear their own voices and participate by saying "Pass."

The direct question is another method, but you must use it with care. Something like the following may be appropriate: "Sandy, you know this process well. Do you have a suggestion?" But as the leader, you have to know your group, and know whether putting someone gently on the spot will help or hinder development.

The Second Pass.

After the initial brainstorm and allowing time for further thinking, it's a good idea to have another session in order to capture additional ideas. These ideas come to mind as the group members think over the problem and consider what was said.

You can handle the second pass in two ways. One way is to gather the group together and go through a second brainstorm with a time limit of 10 or 12 minutes. The same rules apply as in the first pass. The main purpose of this session is to record all the ideas that have come to mind since the first brainstorm.

Another technique for the second pass is to post the brainstorm sheets in the workplace so that people can jot down their ideas as they have them. Posting the sheets has another advantage: it allows people who work in the same area the opportunity to contribute, even though they are not a regular part of the problem-solving group. In this way, they don't feel that they are being left out.

COMPLETING THE BRAINSTORM: A THOROUGH SOAKING

How do you make sure that your brainstorm has covered all possible causes for a problem? Well, you can't. Sometimes you're up against really tough problems. Sometimes the solution lies with an outside expert. Often, though, the solution is right on your own doorstep.

Even if you don't solve the problem right away, you can make sure that you have covered all the general areas of possible causes. Make a list of the general areas and then check whether your group has examined every one of them.

Such a list would include a number of subjects. Earlier in this module, we mentioned the need to solve chronic problems. Major factors to consider when examining chronic problems are method, material, machine or equipment, environment, and employee. *Method* concerns the process you follow when doing your job. What are the steps in performing a CAT scan? What is the procedure in servicing a hotel room? *Materials* are the elements that come to the job, such as data, forms, linens for hotel or hospital rooms, sales reports coming to a clerical person for word processing. *Machine* or equipment includes items like typewriters, laser scanners, X-ray equipment, and computers. *Environment* may be a factor. Cleanliness is important in health and food services. Finally, the *employee* is the person doing the job. Factors connected with the employee could include training, level of skill, eyesight, and literacy.

Other general areas may also apply to the problem. Your group might want to consider factors such as management, money, competition, and government regulations.

DIFFICULTIES WITH BRAINSTORMING AND WHAT TO DO ABOUT THEM

You're stepping on my turf!

In a brainstorm, some people may be very close to the problem and some may not. For example, a group of bank tellers may be led by their supervisor, and a loan officer may be sitting in. When an idea comes from the "outsider," the "insiders" may resent it. What do you do?

Training the group is one answer, and it is a must. While training your group in the goals for brainstorming—self-development of each person, solving problems, and development of team spirit—you need to explain that we all have blinders on. Each of us sees only part of the problem and its causes. Also, remind the group that an outsider may see things that insiders can't. That's one reason we need outsiders' views. It's a matter of perspective.

Criticism.

Above all, build a constructive, positive environment in the group. Do this by criticizing problems, not people. Make sure that ideas, not persons, are evaluated. Be especially careful that mistakes are not publicized and never appear in anyone's personnel file.

The difficult member.

Some people are difficult to deal with in groups. They talk too much; they get off track; they criticize people instead of ideas or they shoot down the ideas. How do you handle this kind of person?

You need to be firm but friendly. First, try talking to the person privately. Explain how his or her actions—for example, talking too much or shooting down ideas—disrupt the group's work or put down individual members. Give the difficult person special jobs to do for the group. Sound out his or her feelings about what the group is doing. Don't fight them; draw them in.

At other times, direct confrontation may be necessary. When a person gets the discussion off track, gently direct the conversation back to the topic. In dealing with criticism of people or ideas, remind the group as a whole of the ground rules.

In our experience, difficult people usually do one of two things: they either become your strongest supporters or they leave.

CAUSE AND EFFECT DIAGRAMS: ORGANIZING THE BRAINSTORM

Once you have finished the brainstorm session and your team's ideas are written on a flipchart, you may ask, "What's next?" Very likely, your brainstorm list is a jumble of ideas. It needs to be organized so that you can use it effectively.

At this point you can use a second problem-solving method, the *cause and effect (C and E) diagram.* This diagram shows in picture form how one idea from the brainstorm relates to another.

Figure 2-1 is a cause and effect diagram. In the box at the far right is the "effect," or problem: hindrances in cleaning. Everything to the left of the boxed-in effect is a possible cause of the effect. We emphasize "possible" because at this point, we don't know for certain the real cause or causes.

These possible causes all came from a brainstorm session. In that session, let's suppose that you, as the leader, asked your group members to

Figure 2-1. Cause and effect diagram.

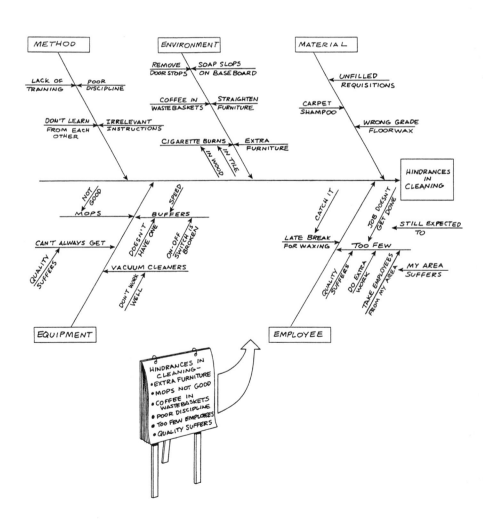

brainstorm possible causes of the hindrances in cleaning. Your group came up with 17 possible causes.

In the C and E diagram, you have organized the brainstorm list of causes under several main headings. Under "material," for example, you listed "unfilled requisitions," "carpet shampoo," and "wrong grade floor wax." These three terms came from the brainstorm list, and all three describe supplies or materials. That's why you drew them as branches off the main arrow, "material."

You will see four other main arrows: method, equipment, environment, and employee. All the other ideas from the brainstorm fit under one of these main arrows. For example, "buffers" and "speed" fit under "equipment" and "coffee in wastebaskets" fits under "environment."

WHY USE THE CAUSE AND EFFECT DIAGRAM?

There are several reasons for using the cause and effect diagram. First, it organizes the ideas of your brainstorm session and helps to sort them into basic categories. Second, it shows relationships between ideas. Buffers may be a source of hindrance in cleaning. But the reason for the problem with buffers is that the on–off switches don't work. So the relationships are as follows:

"on–off switch is broken—buffers—equipment—hindrances in cleaning."

Grouping the ideas under main headings also helps to complete the brainstorm. When you and your team look over the C and E diagram, you may see some gaps that need filling. For example, suppose there were no ideas under "environment." This would suggest that "environment" had been forgotten during the brainstorm session. Finally, the C and E diagram helps your team to keep track of where they are in the problem-solving process. It serves as a record of the brainstorm.

HOW TO CONSTRUCT A CAUSE AND EFFECT DIAGRAM

The C and E diagram is fairly simple to construct. As you and your team work on it, you will see more relationships between various ideas.

Step 1. Gather the materials.

You will need a big flipchart or large sheets of paper, masking tape, water-based flipchart markers with fairly broad points, and the brainstorm idea list.

Module 2: Basic Problem-Solving Tools 35

Step 2. Call together everyone involved with the problem.

Generally, this group will include the leader and the members of the brainstorm group, but it may also include outsiders such as users of the facility or people from building maintenance. One person may volunteer to act as a recorder and draw the diagram.

Step 3. Begin to construct the diagram.

On the right-hand side of the paper, write down the problem or effect. State it clearly so that everyone understands what will be discussed. Next, draw a box around the problem. See Figure 2-2.

Figure 2-2. The problem boxed in.

Step 4. Draw the spine of the "fishbone."

Begin at the left-hand side of the paper and draw an arrow to the box. See Figure 2-3.

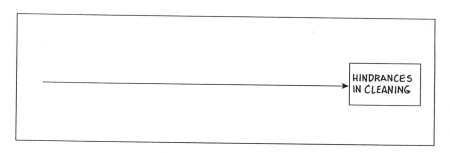

Figure 2-3. The "spine" of the fishbone drawn in.

Step 5. Add the main causes.

"Equipment," "method," "material," and "employee" are the four main headings most often used, but you and your team might decide that others are also appropriate. For instance, as we suggested earlier, you will probably want to add "environment" because many ideas from the brainstorm seem to concern the work environment such as cigarette burns in the tile or wood surfaces, extra furniture in the room, and coffee in the wastebaskets. Other possible main causes might include:

causes outside the process
customer expectations
money
management
government regulations

Your diagram will now begin to look like Figure 2-4.

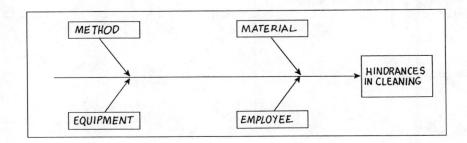

Figure 2-4. Main causes, or "bones," of the fishbone drawn in.

If the team has decided to use additional causes, your C and E diagram will look like Figure 2-5.

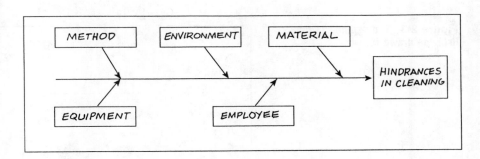

Figure 2-5. Additional main bone.

Step 6. Add the brainstorm ideas.

At this point, begin sorting through your brainstorm ideas to group them logically under the appropriate headings. These ideas may be subdivided further. See Figure 2-6.

Module 2: Basic Problem-Solving Tools 37

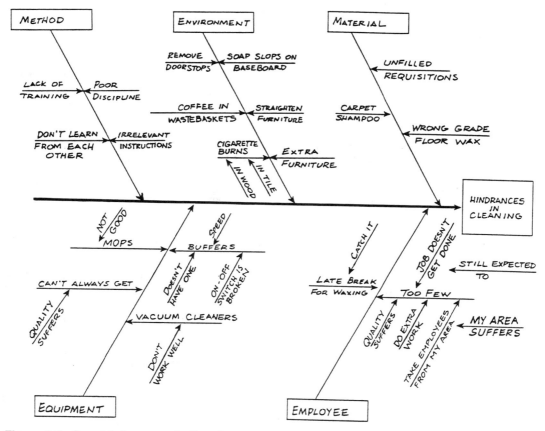

Figure 2-6. Completed cause and effect diagram.

THE PROCESS OF CONSTRUCTING THE CAUSE AND EFFECT DIAGRAM

The ideas for the cause and effect diagram can come from a previous brainstorm session, or you can suggest the ideas as you build the diagram. If you suggest the ideas as you go along, the process of constructing the C and E diagram is like brainstorming. You need a leader to guide the session and someone to serve as a recorder. The recorder works directly on the C and E diagram. The basic rules of brainstorming apply here, too. Make contributions in turn, pass when you don't have an idea, and refrain from criticism.

As in brainstorming, working out the diagram requires effort and guidance from the leader. There are many decisions to make about which ideas go where. The leader may need to ask, "When does this happen?

Why does this take place?" At the same time, the emphasis must always be on how to solve the problem, not on who's to blame.

There are two ways to fill in the ideas under the causes. The team can brainstorm in a freewheeling manner and cover all the headings at once, or work through each cause in turn. If one area does not receive much attention, the team can concentrate on it for a few minutes.

What do you do when an idea seems to fit under more than one heading? Include it under every heading it seems to fit.

TYPES OF CAUSE AND EFFECT DIAGRAMS

The most commonly used C and E diagram is the fishbone diagram, as in Figure 2-1. This diagram organizes and relates the causes of a problem.

Another type of C and E diagram is the *process C and E diagram*, which shows each major step in the sequence of delivering a service. To construct this kind of diagram, the team looks at each stage of the process and determines which main "bones" are involved. See Figure 2-7. Then the group brainstorms each stage for whichever ideas apply to the main bones. In this example, they brainstorm "remove dirty dishes," putting ideas under materials, method, and employee. They do the same for the next step, "wipe up table." In this step, the main bones are employee, method, and equipment. The team may also consider other steps.

Brainstorming and cause and effect diagrams are two methods for finding the causes of service problems. Both methods can be used at several stages in the problem-solving process, such as finding possible causes and developing trial solutions.

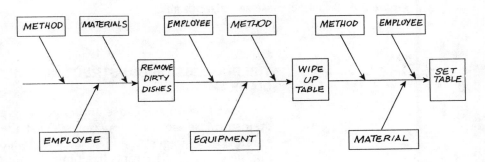

Figure 2-7. Process cause and effect diagram with main bones.

PARETO ANALYSIS

A third method, *Pareto* (pa-RAY-toe) *analysis*, is particularly useful in dealing with chronic problems because it helps you decide which of several chronic problems to attack. You can also make a Pareto analysis at the end of your problem-solving process to see whether your solution worked.

You've probably heard the saying, "It's the squeaking wheel that gets the grease." In choosing problems to solve, it's often the same way. The problems that seem biggest or most urgent are the ones that attract our attention. To be an effective problem solver, however, you must be able to sort out the few really important problems from the more numerous but less important problems. Pareto analysis will help you do this.

You may already know the idea behind Pareto analysis, even though you haven't heard it called by that name. You see it when 20% of your customers account for 80% of your business. To give another example, it's when only a few branches of a bank, out of many branches, conduct most of the business for the bank.

Problems tend to sort themselves out this way, too. When you do some investigating, you usually find that of the 12 or so problems you have looked at, only one or two, maybe three cause the most dollar loss, happen the most often, or account for the most trouble.

Figure 2-8 is a special type of bar graph called a *Pareto diagram*. In this example, a problem-solving team of hotel room attendants has collected data about items for which there were shortages during their weekend cleaning. The team looked at the past month's reports and recorded the data in a Pareto diagram (see Figure 2-8). Among the 68 recorded shortages, they found 33 instances where they didn't have enough sheets; 18 times they were short of pillowcases; towels, 7 times; shampoo, 5 times; toilet tissue, 3 times; and twice they didn't have a vacuum cleaner. Sheets accounted for 33 of the 68 shortages, or 48.5%.

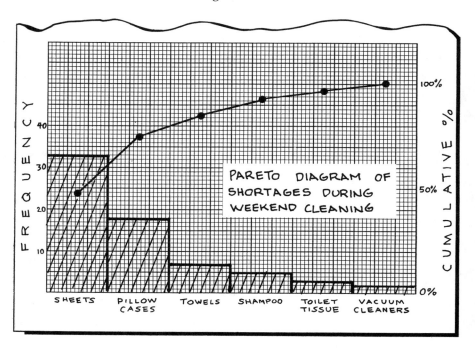

Figure 2-8. Pareto diagram.

If this team wants to improve the availability of supplies and equipment during weekends, which problem do you think they should tackle first, vacuum cleaners or sheets? A missing vacuum cleaner is an obvious problem, but the team discovered that missing sheets was far more frequent. Therefore, to improve both quality and productivity, the best approach to use is to tackle missing sheets, the biggest problem. Pareto analysis helped the team make this decision.

What is the difference between Pareto analysis and a Pareto diagram? As we said before, a Pareto diagram is a special type of bar graph. In this graph, the problem that occurs most frequently is shown by the first vertical bar at the left, the tallest bar. The next most frequent problem is represented by the second tallest bar. The third bar shows the third most frequent problem, and so on. "Frequency" might mean cost in dollars, number of errors (as in invoices), or how often a computer program fails. Each bar represents a specific category, day of the week, or even employee.

Pareto analysis will help you see that there is a need for change and improvement. The Pareto diagram is the picture part of the analysis. Pareto analysis helps set priorities on problems that need to be solved and helps bring agreement on what to do first. It helps you make decisions based on data, not on "squeaking wheels."

HOW TO CONSTRUCT A PARETO DIAGRAM

Step 1. Specify your goal clearly.

Let's suppose your company is a retailer. You are a member of a task force that has been asked to improve the quality of handling customer orders. Your team is not responsible for improving the quality of the actual merchandise itself, but you are asked to look at such things as the following: Did the customer receive the correct product? Was it delivered on time? Was it billed properly? This might require your group to address billing errors, shipping errors, or any of a host of other potential problems. Your team has been instructed to look at quality from the customer's viewpoint. Pareto analysis will help you decide which specific problem area to address first.

Step 2. Collect data.

First determine whether data are already available. If not, then collect the data.

In our example, you know that the accounting department has copies of all credits issued to customers for the last several years. Although there are many ways to determine the customer's view of quality, reviewing credits is a good beginning. One member of your group volunteers to obtain data for the last 6 months from accounting. Sandy's list is shown in Figure 2-9.

Figure 2-9. Listing of credits.

LISTING OF CREDITS

Recorder: Sandy Main   Source: Corporate Accounting
Date: Dec. 5

Customer	Date	Reason	Dollars
Buckman	3/25	Pricing error	25.00
Natural	4/20	Shipped late & out of order	816.00
Natural	6/03	Shipped in error	520.00
Shamers	6/17	—	69.00
Johnstone	3/13	Billing error	20.00
Printographics	5/07	Billing	126.00
Providential	2/02	Late shipment	134.00
Artistic	1/12	Short one carton	6.00
Custom	3/30	Wrong shipment	8.00
Banker	5/15	—	40.00
Niceties	2/23	Shipping shortage	105.00
Natural	2/17	Order entry omitted printing	278.00

Step 3. Tally the data.

Count up the items in each category.

In our example, the group decided to look at "dollars" first, although this is not the only way the data could be analyzed. A tally sheet is shown in Table 2-1.

TABLE 2-1
Tally of data.

Pricing: $25	$25.00
Shipping: $816, 520, 134, 8, 105	1,583.00
No reason given: $69, 40	109.00
Billing: $20, 126	146.00
Short: $6	6.00
Order entry: $278	278.00
Total dollars	$2,147.00

42 SPC Simplified for Services: Practical Tools for Continuous Quality Improvement

Step 4. Rank the categories by size.

List the most frequent category first, whether it's the most dollars, as in our example, or the largest number of defects, or whatever you are looking at. Use the top part of the Pareto Diagram Worksheet (Figure 2-10).

Figure 2-10. Pareto diagram worksheet.

Category	Frequency	Cumulative frequency	Cumulative percentage

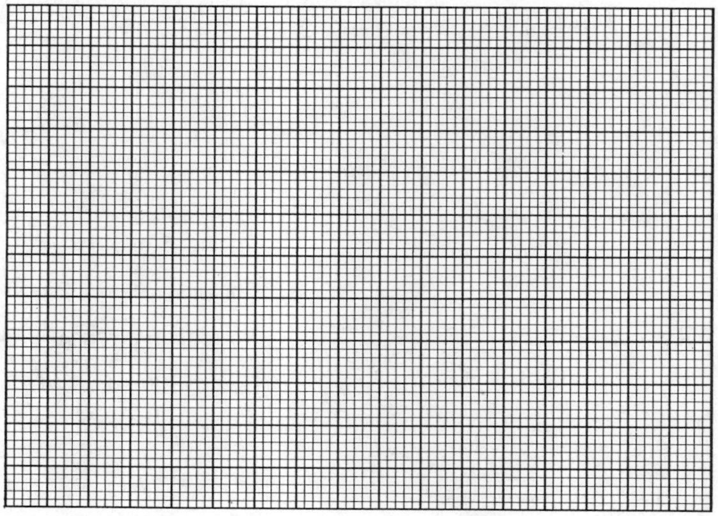

PARETO DIAGRAM – WORKSHEET

When you have rearranged the tally by frequency of dollars, your worksheet will look like Figure 2-11.

You are now ready to organize the data into a Pareto diagram.

Figure 2-11. Tally of data, rearranged in order of frequency.

PARETO DIAGRAM – WORKSHEET

Category	Frequency (Dollars)	Cumulative frequency	Cumulative percentage
SHIPPING	1,583.00		
ORDER ENTRY	278.00		
BILLING	146.00		
NO REASON GIVEN	109.00		
OTHER	31.00		
TOTAL DOLLARS	2,147.00		

Step 5. Prepare the chart for the data.

On the graph portion of the Pareto Diagram Worksheet, draw horizontal and vertical scales. Then mark the numbers on the left-hand vertical scale so that the largest category will fit comfortably. Your largest category is "Shipping," with $1,583.00, so run your vertical scale up to at least $1,600.00. We ran it up to $2,000.00. Label the scale "Dollars." See Figure 2-12.

Next, subdivide the horizontal scale into equal-width intervals so that you have enough intervals for your categories. You may decide to combine the smallest categories into a single group called "Other." We recommend that the total of those combined in "other" not be more than 10% of the overall total.

You have six categories or reasons for credits. The two smallest, "Short" and "Pricing," have very small dollar amounts, so they can be combined into one category—"Other." As a result you will need only five equal-width intervals. See Figure 2-12.

Step 6. Draw in the bars.

For the first interval on the left, draw in a bar to represent your largest category. "Shipping" is the largest category with $1,583.00, so draw it in with a height of $1,583.00 and label it "Shipping."

For the second interval from the left, draw in a bar for "Order Entry," the next largest category, at a height of $278.00. Label it.

Continue in this manner with all the other categories.

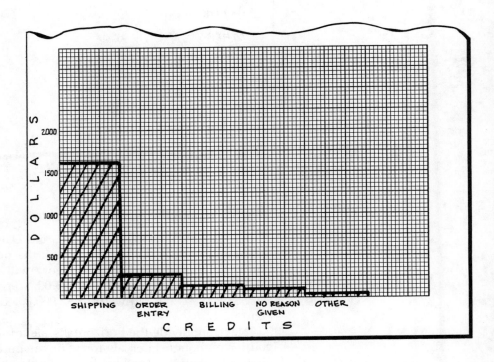

Figure 2-12. Pareto diagram with bars drawn in.

Step 7. Make calculations based on tallies.

On the top half of the worksheet, first add the dollars, starting with $1,583.00 ("Shipping"), the largest. Add the next largest entry, $278.00 ("Order Entry"), for a total of $1,861.00. When you add the dollar amounts for "Shipping" and "Order Entry," the *cumulative frequency,* or amount, is $1,861.00. The third entry, $146.00 ("Billing"), gives a cumulative amount of $2,007.00 because you added "Shipping"—$1,583.00, "Order Entry"—$278.00, and "Billing"—$146.00. Make a note of each cumulative amount and keep adding until you reach the last entry. The last cumulative amount should be $2,147.00, the total of all the credits. See Figure 2-13.

Now that you have calculated the cumulative dollars, the next step is to find the *cumulative percentage* for each entry. To do this you use a simple formula:

Cumulative percentage equals cumulative amount
divided by the total amount times 100%.

As you have already seen, the first cumulative amount is $1,583. Divide $1,583 by $2,147, the total of all credit dollars. The result is 0.74. Multiply this figure by 100% to find the cumulative percentage.

$$1{,}583 \div 2{,}147 = 0.74$$
$$0.74 \times 100\% = 74\%$$

Figure 2-13. Top part of Pareto diagram worksheet showing cumulative dollars and percentages.

PARETO DIAGRAM – WORKSHEET

Category	Frequency (Dollars)	Cumulative frequency	Cumulative percentage
SHIPPING	1,583.00	1,583.00	74
ORDER ENTRY	278.00	1,861.00	87
BILLING	146.00	2,007.00	93
NO REASON GIVEN	109.00	2,116.00	99
OTHER	31.00	2,147.00	100
TOTAL DOLLARS	2,147.00		

To find the second cumulative percentage, take $1,861, the second cumulative amount, and divide by $2,147. The cumulative percentage is higher this time because now you are using the cumulative amounts from *two* categories.

$$1,861 \div 2,147 = 0.87$$
$$0.87 \times 100\% = 87\%$$

Continue in this way to find the cumulative percentage on the basis of each cumulative amount, and make a note of each cumulative percentage. The last one should equal 100%.

Step 8. Complete the Pareto diagram.

Finally, set a scale on the diagram to show the cumulative percentages. You can use 10 major subdivisions of graph paper. We recommend that you (1) use an easy method and (2) use a scale that is easy to read, fits the graph well, and looks good. The following method is simple and clear.

Draw a vertical scale for cumulative percentages on the right-hand side of the graph. Label it "Cumulative %." Mark off 10 divisions to represent 10 percentage points each. Mark some cumulative percentages on it—at least 0%, 50%, and 100%—so that the reader will understand the scale.

Next, draw in the cumulative percentages. At the bar for "Shipping," mark a small dot at a height of 74%, your first cumulative percentage. Over the "Order Entry" bar, draw a dot at 87%, and so on. Then connect the dots with straight lines to make the chart easier to read.

Finally, label the chart "Pareto Diagram of Credits, Jan. 1–June 30." This completes the Pareto diagram. See Figure 2-14.

HOW TO INTERPRET THE PARETO DIAGRAM

Your Pareto diagram (Figure 2-14) is a document of quality problems: it can be duplicated, put on an overhead transparency for a presentation, or otherwise copied and stored.

The Pareto diagram is also a communication tool based on data, and it can help to bring agreement about which problems should be solved first. The diagram allows the process to tell its own story without politics or personal feelings. In Figure 2-14, for example, anyone—management, supervisors, hourly employees—can see that shipping is by far the highest dollar category of credit in the time period you studied and should be tackled before anything else.

Finally, the Pareto diagram serves as a way to compare problems that existed before you worked to improve the process with problems that

Figure 2-14. Completed Pareto diagram of credits.

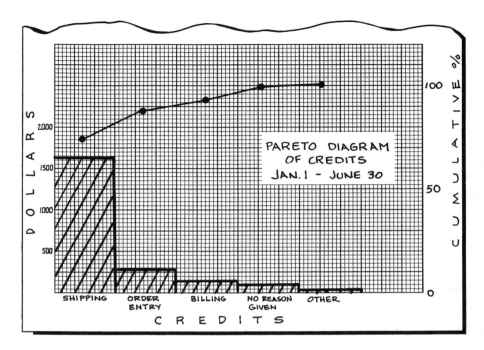

exist after you have worked on the process. After you tackle the problem of shipping, a new Pareto diagram should show shipping as a minor or nonexistent problem.

SUMMARY

Three very powerful tools for improving your process are brainstorming, cause and effect analysis, and Pareto analysis. These three problem-solving tools are especially useful in solving chronic problems.

Brainstorming is a method by which a group can identify problems and gather many suggestions for the causes of the problems. The brainstorming process encourages creativity and participation by everyone in the group. Brainstorming allows room for wild ideas and piggybacking—building on someone else's idea. The proper place for criticism is after the session, and then it should be concerned with ideas, never with people.

The cause and effect diagram organizes brainstorm ideas into categories such as method, material, environment, equipment, and employee. This organization shows how the different brainstorm ideas relate to each other. The C and E diagram completes the brainstorm because it reveals areas that may have been overlooked during the brainstorm. It also helps you keep track of where you are in solving a problem. Any group of people who are solving problems can use the C and E diagram.

The process C and E diagram is useful for tracking a service through the sequence of operations. This type of diagram analyzes each stage of the process.

The last tool, Pareto analysis, helps separate the few important problems from the many unimportant ones. The Pareto diagram is the picture part of Pareto analysis. When completed, it serves as a document of quality problems and as a tool for communication. The Pareto diagram can be the basis for comparing a situation before and after the problem is solved.

PRACTICE PROBLEMS: QUALITY PROBLEM-SOLVING TOOLS

Work through the following problems, using the problem-solving techniques you learned in Module 2. The solutions can be found in the "Solutions" section beginning on page 239.

Problem 2-1a.

Brainstorming requires a group in order to work well. You need other people to make creative ideas flow. If you aren't part of a group, try a brainstorm session with your family or several friends at lunch. Even a 10-minute session can be fun and productive.

If your group is new to brainstorming or if the members are not used to working together, try a fun brainstorm first. This will accustom the members to the rules of brainstorming and it will show them that brainstorming can be both productive and creative. Try one of these topics:

Uses for old books.
Uses for discarded Christmas trees.
Uses for old tires.

It is not unusual for brainstorm groups to think of 30 or more uses for each article. Pick one of these topics as a fun brainstorm and see if your group can beat the record.

Problem 2-1b.

Once your group has had some experience with brainstorming, they will find it easier to use on a work-related problem. Here is a practice problem that could take place anywhere in the service industry.

Your department is developing a training program for employees new to the company. These people will be unfamiliar with company policies, procedures, and possibly the service your company provides. Brainstorm the major issues for which training will be needed.

Problem 2-2.

In a brainstorm session, your team has generated a list of possible causes for disorganized stock in inventory. You have offered to arrange these ideas on a cause and effect diagram. Using the following brainstorm ideas, set up a C and E diagram.

Brainstorm List: Disorganized Inventory

Lack of space	Communication
Bin labeling	Employee friction
Duplication of effort	Out of stock
Too much stock	Obsolete stock
Labels not visible	Wrong labels
New stockroom employee	Different methods
Stock in wrong bins	Label errors
Rush orders that aren't really rush	Missing labels
	Unclear requisitions
Clutter	Different supervisors
Not following procedures	Hot and humid in summer
Procedures not clear	Large cartons of incoming materials
Stuffy air	

What major causes, or "bones," did you use on your C and E diagram? Did the brainstorm group cover all the major areas? If it neglected one area, how can this area be covered? Are there any other possible main causes that your group might consider for this problem?

Problem 2-3.

Your office manager is concerned about possible ill will from customers caused by the data entry problems your department is experiencing. You are part of a task team that has collected four weeks of data. The report follows:

Data Entry Errors

Misspelling	43
Wrong price	17
Wrong quantity	24
Wrong customer address	11
Unclear handwriting	31
Reversed digits	7
Incorrect catalog number	58
Price change	47
Incorrect description	13

Your manager is very familiar with the problem-solving tools that you have been studying. You are asked to use those tools in determining where are the best opportunities for reducing data entry errors. Make a Pareto diagram of the errors. How will you set up the horizontal and vertical axes? Can you combine any of the categories? Why or why not? On the basis of the diagram, what recommendations could you make to your manager?

MODULE 3
Quality Improvement Tools

NEW TERMS IN MODULE 3 (in order of appearance)	
process flow chart	*storyboarding*
operation	*pinner*
move	*scatter diagram*
inspection	*variable data*
delay	*dependent variable*
storage	*corner count test*
decision	*straight-line relationship*
connector	*median*

It doesn't matter whether you work in banking, food service, government, the nonprofit sector, travel, transportation, or any other service industry. Quality is important. It is important to your customer; it is important to your competition; and it needs to be important to you. The tools in this module do more than simply help identify sources of problems; they help you start improving the quality of your service and your process.

FLOW CHARTS

Have you ever asked: "Why does it take so long to get a simple voucher approved?" or "Why do we have so many steps to follow in processing this type of policy? Isn't there a simpler way?" A tool to help you answer such questions is the *process flow chart*.

The process flow chart is a special kind of diagram that pictures the steps of a particular job in sequence. This diagram helps track the flow of information, paper, material, or people through the system of delivering a service. The flow chart may show that the system is more complex than anyone realizes. If you or your problem-solving team can see how the material or paper or person moves through the system, you may be able to come up with a simpler way. There may be repeated or unnecessary steps. Once you know the actual steps in the process you are studying, you probably can find ways to combine or eliminate unnecessary ones. Simplifying the system that delivers the service is a good way to begin improving quality, efficiency, and productivity.

An important use of the flow chart is helping to identify the points in the system that need to be controlled. One or more steps may be critical to providing the service on time or in an effective manner. Or you may discover where in your process trouble usually happens. Such places are ones you will definitely need to control. The flow chart helps you find the most effective point to control. (Modules 5 and 6 discuss methods for controlling a process.)

PROCESS FLOW-CHART SYMBOLS

A process flow diagram is fairly easy to construct. But first, you need to know the flow-chart symbols.

Operation: This is the work that's required to complete a task. You are doing something such as filling prescriptions, creating data files, changing reservations, folding towels, or adding new items to the menu.

Move: Something—information, a person, paper, supplies—travels from one point to another. Cleaning supplies go from central receiving to the storage area on each floor. A technician takes blood samples to the lab for analysis. Daily receipts go to the front office for tabulation.

Inspection: Someone tests or verifies that the material, information, form, or activity is correct and meets the requirements. This person decides whether the material, form, or activity should continue to the next step or if a correction, addition, or some other change is needed.

Delay: A delay means waiting; for some reason, you can't go immediately to the next step in the process. A delay can occur before an operation, an inspection, or a move, as well as after. For example, the drive-through teller has to wait for the balance inquiry before sending the customer her cash. Frozen foods sit in the supermarket aisles waiting to be stocked in the freezer.

Storage: This is a holding area. Computer invoices sit in an in basket until the billing clerk checks catalog prices. Housekeeping stacks clean linens on carts after folding. Policies are stored on microfilm.

Decision: This symbol is sometimes used to indicate a place where alternative actions may be taken. For example, if insurance carrier A is listed, complete form 1; if insurance carrier B is listed, complete form 2.

Connector: This symbol indicates a continuation of the process. If you run out of room on a page, place a connector symbol at the end of the line and at the beginning of the next.

CONSTRUCTING A PROCESS FLOW CHART

Step 1. Define the process.

You may think defining your process is a lot like trying to eat a moose! Where do you start? But it really isn't too difficult if you identify the beginning and end of the process you want to study. Ask yourself, "What is the first thing I do?" Make a note of this step. Then ask, "What is the last step?" and note that. For the moment, don't be concerned with the activities between the beginning and the end points. If you think of an earlier step, add it. Approaching the study of your process in this way is not overwhelming. You can begin to "eat the moose"!

Step 2. Identify the steps in the process.

The easiest way to identify the steps is mentally to "walk through" the process as it normally happens. In this way, you will add the steps between the beginning and end of the process. A new step begins when a new kind of activity is required.

Write down the steps on a sheet of paper. You may actually need to go where the activity happens, observe, and take notes. Other methods you can use to define your process steps are "storyboarding" (explained later in this module) and making a video tape of the process. Be sure you include every operation, move, point of inspection, storage area, and delay. It's important to list all the elements of the process regardless of how long or short the time it takes to complete each one.

Step 3. Draw the flow chart.

Once you have defined the process and have identified all the steps, you are ready to draw the flow chart. Choose the correct flow-chart symbol for each step and draw that symbol on a sheet of paper, flipchart, or chalk board. Briefly identify each step, telling who, what, or where.

Connect the steps with a line. As you are putting the diagram together, the line may become too long for the page. Use the connector symbol and start another line. Continue in this manner until you have covered all the steps in the process.

Sometimes the flow chart may branch. See Figure 3-1. This can happen in a number of ways. Sometimes several operations need to happen at the same time. For example, two trainers are creating a training package and each is developing one part of the materials. The chart can also branch during an inspection step. A supervisor checks for typing errors in policies and those with errors are returned for correction. See Figure 3-2. Another type of branch results from an "if," or conditional, situation. If a voucher is below a certain amount, it goes directly to the bursar for payment. If it is above this amount, someone takes the voucher to accounting for an additional signature. See Figure 3-3.

Figure 3-1. Branch in flow chart showing parallel operations.

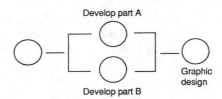

Figure 3-2. Branch in flow chart showing correction loop.

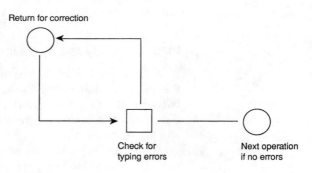

Figure 3-3. Branch in flow chart resulting from an "if."

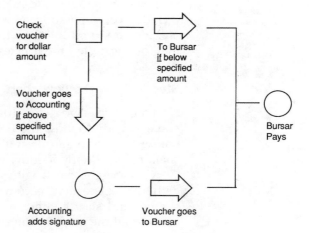

Step 4. Determine the time or distance for each step.

In order to improve the process flow, it is important to know the time it takes to complete each step. This helps you find where you can reduce or eliminate wasted time.

When determining how long a step takes, note its start and finish. Then time it. You may already have an idea of the time based on your own experience. Another thing to try is actually tracking a document, material, voucher, or person through the process. If the time it takes to complete a step varies a lot, you may need to record times for a week or a month to determine the step's average completion time. This could happen when a bank teller verifies night deposits. The job will probably take longer to complete on Mondays than it will on Thursdays.

Write the appropriate times beneath each step. Remember to record the time for delays and storage. These are good places in the process flow to target for improvement.

Also you *may* want to record the distance for each move. Possibly you already know how far it is from one area to another. A move can be one area where you can improve your process.

Step 5. Assign a cost for each step.

You may want to assign a cost to each step in the process, but this is optional and will depend on your particular situation. Cost information could give incentive to eliminate unnecessary or duplicate steps.

HOW TO USE THE PROCESS FLOW CHART

Once you have drawn the flow chart and have determined times, distances, and costs for the different steps, you can begin to use it in one of two ways: to control the process or to improve it. The flow chart helps you decide what steps need to be controlled and where the overall process requires improving.

Controlling the process.

As you analyze the flow chart, you may find that many of the steps seem to be working well. For the time being, these steps should remain unchanged—but they need to be monitored or controlled so that change does not occur. The flow chart can also help you find the points in the process that are giving you trouble, that is, unwanted change. Once you identify such points, consider using a process cause and effect diagram to examine the elements of your process steps. By looking at the method, the equipment, material, or people used in the process steps, you may be able to find causes of the problem.

Trouble points probably require some type of control chart to help keep the process performing as you want. In Modules 5 and 6, you will learn about control charts. These are statistical tools that help you control processes.

Improving the process.

Improving the process means that you will deliberately change it in some way. Can you eliminate any repeated operations? Are there ways to shorten or eliminate delays and moves? How can you shorten the time that things like information, paper, or supplies are in storage? Consider using brainstorming, process cause and effect diagrams, or storyboarding to improve your process.

Here are some other points to keep in mind.

1. Is there a point in the process that slows or restricts the flow of work, information, or people? What can we do to improve this situation?
2. How can we improve the sequence of the operations to make the process more effective? Would a change in the workplace or the people increase effectiveness?
3. Can we improve how we do the operation or activity?
4. Can we reduce or eliminate having to correct, change, add, or recycle something in the process?
5. Is there a better way?

Figure 3-4 is a process flow chart of part of a hospital medication delivery system. Numbers in the symbols refer to the steps of the process.

The flow chart in Figure 3-4 shows some potential trouble areas. Two delays, "consultation with doctor" and waiting for "out-of-stock drugs," are points where some statistical analysis might help. For example, how often do these two types of delays occur? Another critical point is step 6. Statistical monitoring could be helpful here. Do you see other points that are critical to the safe and efficient dispensing of medications? Are there areas where this process could be improved?

STORYBOARDING

Storyboarding is an effective method to use for improving processes, developing ideas, finding causes of problems, and planning projects. It uses some of the principles of both brainstorming and the cause and effect diagram. Like brainstorming, storyboarding is a group process that encourages participation, creativity, and trust. Members generate ideas and organize them under main headings as when developing a cause and

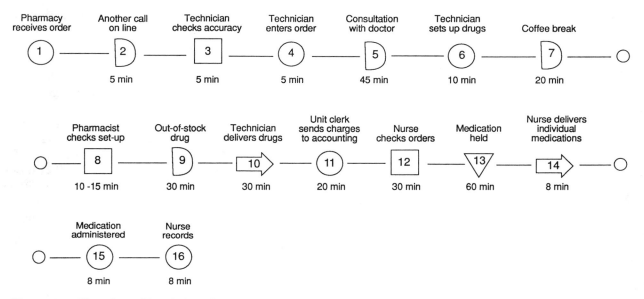

Figure 3-4. Flow chart of hospital medication delivery system.

effect diagram. Identifying all the components of the area you are storyboarding helps later when making task assignments.

Suppose a team is designing a new training program. First, they determine the purpose of the training program. Then they break down the topic into major categories. These categories could include specific training concepts as well as various related issues. Next, members of the team write their ideas for each category on cards and post the cards on the storyboard under the appropriate headings. Someone may act as a "pinner" and do it for them. The result of the storyboard session is a list of the topics where training is needed as well as a list of issues related to training. To complete the storyboard, the group spends time analyzing and selecting ideas so that the project will be manageable.

WHAT IS NEEDED FOR STORYBOARDING?

A group willing to work together.

For a successful storyboard session, we need a group of people who are willing to share their ideas. Who should be included? As in brainstorming, we include everyone who will be a part of the project. There are two reasons for this. First, we need the support of the people who will be a part of the project. Involving them in the development through storyboarding helps win their support for the implementation. Second, we need their ideas, especially to make sure that we don't leave anything out.

A leader.

Anyone can lead the storyboarding: one of the regular members of the group, or even an "outsider."

It's the leader's responsibility to help the team to storyboard efficiently. The leader helps the group determine the type of storyboard session—whether they will be planning, evaluating, or generating ideas. He or she guides the process to make sure that it stays on target, that members of the group have the materials they need, and that people's ideas get placed under the proper headings. The leader exercises enough control to keep things working in an orderly manner. At the same time, he or she encourages people's ideas and participation. He or she must put aside personal goals and serve the group. The leader directs but must also encourage participation.

A pinner.

Depending on the size of the group and the layout of the room, you may need a *pinner*. This person takes the cards as members fill them out and posts them on the storyboard. If the group is small and all members are close to the storyboard, it may work for the members to place their own cards. The pinner should have the chance to contribute ideas, too. Another member can write out ideas for him or her.

A meeting place.

The group should have a meeting place where it will not be interrupted or distracted. This may be a training room, a section of the cafeteria, a supervisor's office, or a conference room.

Equipment.

The group needs chairs, tables to work on, blank cards, water-based markers, masking tape or pins, and a very large board. It helps the group process if the tables are set up so that people can see one another. The cards should measure approximately five by eight inches. Some groups use different colors for header and idea cards. Markers should have broad points so that everybody can easily read the cards when they are posted on the board. The "board" can be a wall where the cards are temporarily attached with masking tape, or a cork board on which cards are pinned.

HOW DOES STORYBOARDING WORK?

The following are some ground rules for storyboarding:

1. Decide what kind of storyboard session this is. Is the group looking for causes of a problem or coming up with a remedy? Are we generating new ideas or are we evaluating ideas from an earlier session? Are we planning a project or developing an implementation?
2. Be sure that all members understand the subject of the storyboard. Post the subject on its own card at the top of the story board.
3. Determine the purpose of the storyboard. Why are we working on *this* subject? What do we want to see happen?
4. Brainstorm for possible categories, or headers, of the subject, using the rules for brainstorming. (See Module 2.)
5. As a group, select main categories from the headers that were brainstormed. Write them on blank cards. Arrange them across the top of the board under the subject of the session. Be sure there is plenty of room for the idea cards under each heading.
6. Work through a heading before going on to the next.
7. As in brainstorming, members write down any ideas they have about the heading on blank cards. They must print in large letters and be brief. Don't worry about spelling.
8. When a card is completed, the member calls the pinner over and tells him or her under which heading the card fits. (If a card belongs under more than one category, write out a duplicate card but wait until that heading is being discussed before pinning the card on the board.)
9. The pinner posts the completed card under the appropriate category as instructed.
10. As in brainstorming, hold all evaluation until the storyboard is done.

The leader monitors the storyboard process and can stop it at any point. If a major category has only a few cards, the leader should direct the group's attention there. If someone comes up with a possible new heading, he or she should ask the group if it should be included. The leader or a member may ask to have a card clarified.

Just as there are different kinds of storyboards, so there are different ways to come up with the ideas under the headings. You can "freewheel it," where each member writes out a batch of cards for the category on which the group is working. Or follow a "round robin" brainstorm, where the leader asks each member in turn for a contribution. In this case, members can write down their ideas at any time, but the pinner picks up only one card at a time.

The following example illustrates the storyboard process for generating ideas. This group has four people in it: Austin, the leader; Amy; Jean; and Mike. Since it is a small group, each person posts his or her own cards.

Austin: Our task is to design a new kitchen for our wilderness camp for inner-city teenagers. A good way to begin this would be to storyboard it. In the first stage of storyboarding, we'll determine our purpose and brainstorm for major categories. Is this O.K. with everybody?

The group agrees, so they begin by storyboarding the purpose of the group's designing the new kitchen. Under the purpose heading are such cards as "to make food service more efficient," "to make it easier to keep clean," and "to meet Trustees' goal of training kids for jobs in food service."

Then the group brainstorms for major categories. After a short session, Austin, the leader, stops them. The brainstorm list follows.

Building codes

Stoves

Equipment

Workspace

Storage

Nonskid floor

Food preparation

Cooking utensils & storage

Austin: Now we select the major category headings from our list. Do you see any duplicates? Should we reword any?

Jean: "Workspace" could be broadened to "Layout."

Mike: It seems to me that "Stoves" really belongs under "Equipment."

Amy: We only scratched the surface with "Building codes." Why don't we change that to "Codes"?

Mike: I think we could include "Nonskid floor" under "Layout."

The categories are printed on blank cards and Mike posts them on the wall. Now the storyboard looks like Figure 3-5.

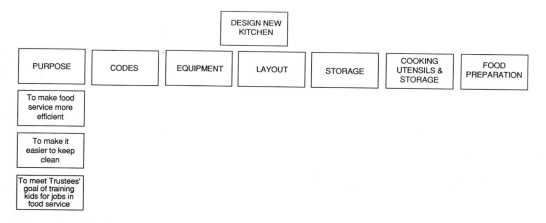

Figure 3-5. Storyboard with purpose and main headings.

Austin: Now that we have our major categories, let's continue storyboarding. Any questions?

Jean: Would you mind going over how we are supposed to do this?

Austin: Sure! In this stage of storyboarding, we work through one heading at a time. Write out your ideas on blank cards. Be sure to print in large letters. Since our group is small we can each post our cards as we write them. As in brainstorming, we hold all evaluations until the evaluation session. If you think a card goes under more than one heading, simply write a second one and post it under both. We'll talk about "fit" later. If you think you have a new category, let the group know and we'll discuss it right then. O.K.?

Jean agrees and the storyboard session continues. At one point, Mike says that "Serving utensils" should be a new category. Austin leads a brief group discussion and the group decides to post it as a new heading but with the title: "Serving utensils & storage."

The storyboard now looks like Figure 3-6.

EVALUATING THE STORYBOARD

In reviewing the cards in our example, we can see several things. "Paper products" appears under more than one heading: "Storage" as well as "Serving utensils & storage." "Food preparation" is completely empty. Maybe the group missed it; maybe it should not be a heading. Over time, additional topics can be added to the storyboard.

The group now has a good beginning on the design of the new kitchen.

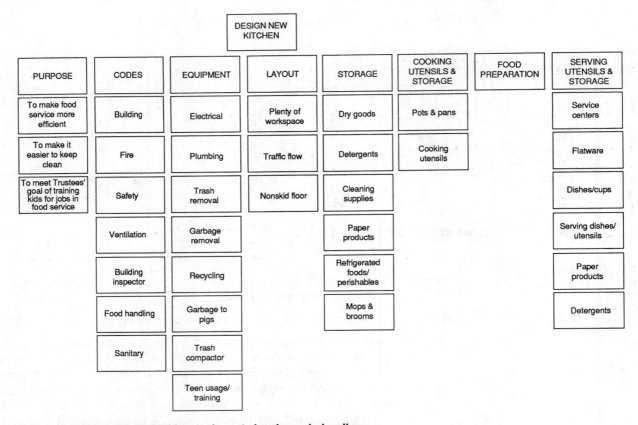

Figure 3-6. Storyboard with idea cards posted under main headings.

The storyboard gives them a number of issues that they must address in their design. Once an item is on the list, they can make sure that it is carefully considered in the design. The list can be shown to others for their input. During the evaluation session, the group selects the most appropriate items, prioritizes them, and makes assignments.

SCATTER DIAGRAMS

Another quality tool for problem solving is the *scatter diagram*. Let's start with an example.

A team of custodians and their supervisors suspected that the higher the humidity, the longer it would take floor wax to dry. To settle the issue, they recorded data on the time it took for freshly waxed floors to dry. Along with the drying times, they recorded the humidity. Then they plotted their data on the chart shown in Figure 3-7. As you look at Figure 3-7, do you see a relationship between drying time and humidity? Doesn't it look as though the longer drying times usually occur during periods of high humidity?

A scatter diagram is a kind of graph that allows us to see how two variables may be related. By plotting our data on a scatter diagram, we can see whether there is a relationship between the variables. If there is a relationship, then by controlling one variable we may be able to control the other. In the drying time and humidity example, if we can control the humidity then we may have more control over the drying time. Even if we cannot control the humidity, we may have a good idea of how long drying will take simply from knowing the degree of humidity.

WHERE TO USE A SCATTER DIAGRAM

Scatter diagrams are useful when you are trying to determine how one variable relates to another. In our example, we are comparing drying time to humidity. There are other situations where you could use the scatter diagram. In a hospital, you might relate the strength of an antiseptic to bacteria count. In the transport industry, you might want to see how the length of time between maintenance of trucks compares to the frequency of breakdowns. In a secondary school, you could relate the number of hours of remedial study to scores on proficiency exams. These examples help you see how scatter diagrams can apply to your work.

HOW TO CONSTRUCT A SCATTER DIAGRAM

The purpose of the scatter diagram is to show whether there is a relationship between two variables. In our example, the custodians and

Figure 3-7. Scatter diagram of floor wax drying time versus humidity.

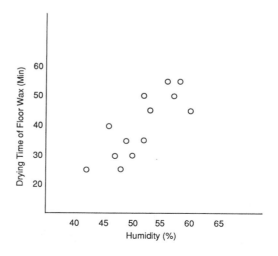

managers think that the higher the humidity, the longer the drying time will be. In other words, they suspect that there is a relationship between the two variables of humidity and drying time.

A scatter diagram is a graph where each plotted point represents two variables. If you look at the drying time and humidity example (Figure 3-7), any one point on the graph represents a particular drying time as well as a specific value of humidity. For example, the point in the lower left corner represents a humidity of 42 percent and a drying time of 25 minutes. This means that when the floors were waxed that time, they took 25 minutes to dry and the humidity was 42%.

What kind of data do you use in scatter diagrams? It should be variable data. *Variable data* come from things you can measure, such as time, length, and temperature. You can measure time to the hour, to the minute, or, if necessary, to the second. Time is a variable. Length, too, can be measured to the foot, inch, or fraction of an inch. Temperature is another variable. Variable data are used in scatter diagrams.

Step 1. Is this a scatter diagram kind of problem?

Check whether you really have the right kind of problem for a scatter diagram. First, does each point you're going to plot come from variable data? Second, are there two things you can measure? That is, are there two variables? Third, are you trying to see if there is a relationship between two variables?

Step 2. Collect your data.

Sometimes the data you need are already recorded. Other times you will have to collect them. When collecting data, record any interesting or peculiar things that happen. For instance, if we noticed that different kinds of wax were being used when we were collecting data for the floor wax study, we should record the kind of wax for each data point. This extra information can be useful later on.

Step 3. Determine the scales for the graph and plot the data.

Start with the vertical axis and decide which variable it will represent. Usually, we plot the *dependent variable* on the vertical axis.

What do we mean by a dependent variable? In our previous example, the different drying times probably resulted from different levels of humidity. That is, the drying times differ depending upon the humidity. When one variable results from or depends upon another variable, this kind of variable is called the dependent variable. In our example, the vertical axis represents the dependent variable "drying time."

Now find the largest and smallest values of the dependent variable, the one you will plot on the vertical axis. Set the scale for the axis so that the largest and smallest values fit inside, *not on,* the edges of the chart. In our example, the longest drying time was 55 minutes, and the shortest was 25. We ran the vertical axis from 20 to 60 minutes.

Use scales that are easy to work with. We marked the vertical scale with the numbers 20, 30, 40, 50, and 60. This makes the diagram easy to construct and to read.

Next, find the largest and smallest values for the other variable. Set the scales of the horizontal axis so that the largest and smallest values of that variable fit inside, but not on, the edges of the chart.

Now plot the data.

Step 4. Do the corner count test.

By doing the *corner count test* a few times, you'll develop the ability to tell at a glance (1) whether a scatter diagram obviously shows a "straight-line relationship" between the variables; (2) whether it's quite clear from the diagram that such a relationship does not exist; or (3) when you must do the corner count test because it's not so clear whether there is a relationship or not. Without doing this step, you could work for a long time without knowing the true situation. That would be a terrible waste of time. The reason for learning this step is to help you become an effective problem solver.

This simple test helps you determine whether there is a straight-line relationship between the variables or if there is only a jumble of plotted points with no real relationship. If a relationship does exist between the variables, it may possibly be a cause and effect one. If there is a true cause and effect relationship, you can predict one variable by knowing the other. Furthermore, if there is a cause and effect relationship *and* you are able to control one variable, you will be able to control the other.

What is a *straight-line relationship?* The ideal straight-line relationship is one where an increase in one variable results in an increase or decrease in the other variable, and the amount of increase or decrease is always the same. A good example of this is shown in Figure 3-8, where we plotted temperature in degrees Centigrade versus temperature in degrees Fahrenheit. For every degree increase in Centigrade temperature, the Fahrenheit temperature always goes up 1.8 degrees. This example is ideal in that all the points fall exactly in a straight line, since degrees Fahrenheit and Centigrade were set up this way. Usually, the data you plot will not fit as neatly as this and you will have a situation more like the scatter diagram for drying time versus humidity.

Figure 3-8. Scatter diagram of degrees Centigrade versus degrees Farenheit.

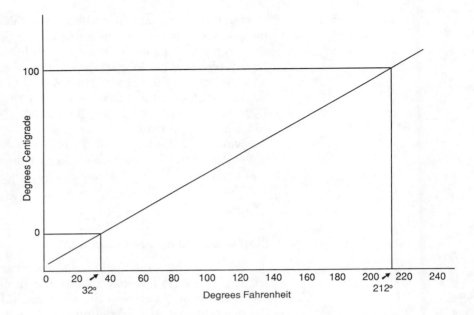

With this introduction in mind, here are the steps to follow to do a corner count test.

A. Do you have enough data?

Are there at least 10 points on the scatter diagram? You need at least 10 for the corner count test to work. If you have less than 10, you will need more help than this book provides. In our example in Figure 3-7, we have 13 points, so there are enough.

B. Find the medians.

Find the *median* for the first variable and draw a line so that half the points are below it and half are above. (The median is where half the points are above the line and half are below.)

To do this, set a ruler on the bottom of the chart so that all the points are above it. Move the ruler up slowly. As you move it up, count the points as they disappear beneath the ruler. Stop when you reach half the count.

If the number of points is odd, 13 as in our example, divide the count in half. This gives you 6.5 plot points. Add 0.5 to round it off—6.5 plus 0.5 gives you 7. Stop your ruler on the scatter diagram and draw a horizontal line through the seventh point. In our example, the line goes through the point representing "drying time of 40 minutes." Six points fall below the line, and six are above it. This horizontal line is the median for "drying times," the vertical variable.

When the number of points is even, say, 38, then divide the count in half (38 divided by 2 is 19). Move your ruler until you have covered half, or 19, of the points. Now position the ruler between the nineteenth point and the next one. The horizontal line you draw across the diagram is the median line because half the points lie above it and the other half are below.

Once you have drawn the horizontal median, find the vertical one. Proceed in the same way, but start your ruler at the right-hand vertical edge of the scatter diagram. Move the ruler in from the right side of the chart. Stop the ruler when you reach half the count. Then draw a vertical line to represent the median. See Figure 3-9 for "drying time and humidity."

Once you have drawn these lines, you will find that half the points are below the horizontal line, and half are above. The vertical line represents the median for the horizontal variable—half the points will be to the left of the line and half to the right. Sometimes you will find your median line has more than one point on it. That's O.K.

C. Label each corner of the chart.

The two median lines have divided the scatter diagram into four parts, or quarters. Mark the upper right one as a "+" or write in the word "plus." Mark the upper left as a "−" or write in the word "minus." Mark the lower left as "+" and the lower right as "−". See Figure 3-9.

D. Do the corner counts.

This is the tricky part, so be careful! Place your ruler vertically on the right side of the chart. Slowly move it toward the left side of the chart.

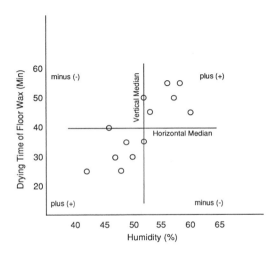

Figure 3-9. Scatter diagram of drying times versus humidity with median lines drawn in and quarters labeled as plus or minus.

Stop at the very first point you come to and note the sign (either plus or minus) of the quarter where the first point lies. Then slowly move the ruler toward the left. If the next point is in the *same* quarter as the first point, count the point. Continue in this fashion, counting the points as they disappear under your ruler. Move your ruler until you come to a point that is in a *different* quarter or is on a *median* line. Stop and write down the number of points you've counted so far. Don't count the stopping point (the one on the median or in a different quarter). In front of this number, write the sign of the quarter where you started counting.

In our example, as we move our ruler in from the right, we find one point in the upper right quarter. It's in a plus quarter. The second point is also in the first quarter. We continue counting until we come to the sixth point, which is on a median line. We stop at a count of 5. Since our first point is in a "plus" quarter, we assign a plus to this count. The first count is +5.

Now place your ruler horizontally at the top of the scatter diagram and move it toward the bottom of the chart. Count as before. In our example, we count two points in the upper right quarter. As we move our ruler further down, we meet two points—both having the same drying time. But one is on the vertical median. So, we stop counting at 2. Since the points we've counted are in a "plus" quarter, we write +2.

Next, count in the same way, coming in from the left side of the diagram. Then count up from the bottom of the scatter diagram toward the top.

Sometimes you will count a point twice. Some of the points you won't count at all. This is O.K.

E. Total the corner counts.

Add all your counts. If some are pluses and some minuses, subtract the minus values from the plus values. The results for our example appear in Table 3-1.

TABLE 3-1

Count from right	+5
from top	+2
from left	+1
from bottom	+4
Total	+12

F. Compare the total of the counts.

Once you have totalled the counts, remove the sign from the total. If it is +12, as in our example, write 12. If the total is −12, simply write 12. The important thing is how large the total is, not whether it is plus or minus.

Now compare your result to the number "11." The number "11" is a figure that statisticians have worked out as a comparison figure for scatter diagrams (as long as there are at least 10 data points). If your result is 11 or higher, you probably have a straight-line relationship between your two variables. But if your result is smaller than 11, there may not be a straight-line relationship. Your data simply don't tell.

In our example, the total is 12, so we can say there is a straight-line relationship between the two variables. The custodians and supervisors have statistical evidence that there is a straight-line relationship between drying times and humidity. The data back up their opinion that the higher the humidity, the longer it takes floor wax to dry.

SUMMARY

The tools in this module can help you improve the process of delivering a service.

The process flow chart is a special diagram that tracks the flow of work, information, paper, materials, or people through the system of delivering a service. The diagram pictures the sequence of steps in a job. Once you have characterized each step as an operation, move, inspection, delay, storage, or decision, you can use the flow chart to analyze the process in order to make it more efficient and effective. Another important use is helping to pinpoint trouble areas in the process that may need to be monitored or controlled with a control chart. This is a tool an individual or a problem-solving team can use to make improvements.

Storyboarding is an effective way to improve processes, develop ideas, find causes of problems, and plan projects. It uses some of the principles of both brainstorming and the cause and effect diagram. Storyboarding is a group process that encourages participation, creativity, and trust. Members generate ideas and organize them under main headings. Follow the rules for brainstorming, such as having a leader and holding off evaluations until the evaluation stage.

First, determine the purpose of the storyboard. Then brainstorm for major categories or headings. Next, participants in the session write their ideas for each category on cards that they or a pinner post under the appropriate headings. The result of the storyboard session is a list of topics and issues related to the storyboard subject. To complete the storyboard, the group spends time analyzing and selecting ideas so that the project will be manageable.

The scatter diagram is a type of graph that helps us determine how one variable may relate to another. Each point on the diagram represents the two variables you are studying. The corner count test is a simple test to see whether, in fact, there is a straight-line relationship between the variables. If the corner test results in a count equal to or greater than "11," there is probably a straight-line relationship, and a cause and effect relationship may exist between the two variables.

PRACTICE PROBLEMS: QUALITY IMPROVEMENT TOOLS

Work through the following problems using the problem-solving techniques you learned in Module 3. The solutions to these problems can be found in the "Solutions" section beginning on page 243.

Problem 3-1.

The billing office manager has asked you to do a flow chart of the morning's activities in your area. You are familiar with the steps, but you make notes for a few days in order to establish completion times for each step in the process. Using the list of steps and times that follow, construct a process flow chart.

1. Arrive at desk where computer printout is waiting.
2. Turn on adding machine. (0.5 min)
3. Begin to sort printout. (30 min)
4. Make four stacks for processing. (15 min)
5. Carry three stacks to desks for other clerks to process. (5 min)
6. Check whether other clerks have all the information they need to start. (5 min)
7. If they are missing pieces for their work, look around for them. (5 to 10 min)
8. Call data processing. If missing pieces are there, go and get them. (8 min)
9. Return to desk. (5 min)
10. May have to wait for invoice batches. (10 min)
11. Begin to verify data on invoice batches with printout. (30 min)
12. Record any corrections on data processing adjustment form. (15 min)
13. Coffee break. (15 min)
14. Record on own register verifications of processing and file batch control slips. (12 min)
15. Balance printouts for two departments, 15 minutes for department A, 22 minutes for department B. (37 min)
16. (But, report for department B is always late.) (8 min)
17. Find reasons for any imbalances and record on printout. (10 min)
18. Make necessary corrections on adjustment forms. (10 min)
19. Check department reports. (8 min)
20. File printout. (6 min)
21. Take adjustment forms to data processing desk. (8 min)

Problem 3-2.

A suburban church wants to become involved in helping the homeless in the local area. They are thinking about providing overnight housing for people needing a place to stay. A task team from the trustees, missions committee, and several other committed people have agreed to do some initial planning. You are part of the team and suggest storyboarding as a good way to begin. For this exercise, set up a storyboard. What are some purposes this team may have for the church's involvement? What are possible main categories? These could be major issues that this team needs to consider. Using these categories, storyboard and fill in these areas.

Problem 3-3.

An auto leasing agency has kept records for two months on the amount of gasoline used in a new model of car in their fleet. For each car, they also recorded the number of miles driven.

Number of miles driven	Number of gallons used	Number of miles driven	Number of gallons used
1523	48.6	1453	61.2
1723	46.8	1980	48.0
2723	78.7	2452	70.8
2602	72.7	927	25.1
1805	49.2	2012	52.7
510	18.2	2303	71.5
762	23.1	1302	37.1
3212	86.2		

(a) Plot these data as a scatter diagram. (b) Use the corner count test to determine whether there is a straight-line relationship between the variables "number of miles driven" and "number of gallons used." (c) Regardless of the result of the corner count test, would you expect a straight-line relationship? (d) Are there any data points that you think might be the result of an error? (e) Are there any cars that may need maintenance?

MODULE 4
Frequency Histograms and Checksheets

> **NEW TERMS IN MODULE 4**
> (in order of appearance)
>
> *frequency histogram* *midpoints*
> *intervals or class intervals* *underlying frequency distribution*
> *boundaries* *checksheet*

WHAT IS VARIATION?

Suppose you and a friend go to the rifle range. Your friend loads, aims, fires, and hits the bull's-eye. Would you conclude, on the basis of this one shot, that your friend is a sharpshooter, perhaps even Olympics material? Or would you ask to see more shots?

Suppose you ask a travel agent for the temperature at a resort area where you want to spend your vacation. The agent says, "75 degrees." Does that one figure tell you enough, or would you ask for a series of temperature readings over a period of time?

Suppose you want to know how long it takes to go through the drive-through window at the local fast food restaurant. Would you be satisfied with a single time, say, 8 minutes?

We think you will say "no" to each of these questions. The rifle marksman may have happened to hit the bull's-eye on the very first shot, and might not hit another in the next hour of shooting. For the resort, you need to know whether 75 degrees is the temperature for summer or winter, night or day. Wouldn't you want to know if the drive-through time was for noon rush hour or another time? A single measurement is not enough.

Why not? Because, as we said before, no two items are exactly the same. They vary. We need a number of examples so that we can tell how good the marksman is; what the temperatures are at the resort area; what the times are at the fast food restaurant.

Variation is natural. It is found in many, if not all, processes. Even the best marksman sometimes misses the bull's-eye; temperatures vary, even at a resort; times to pick up your meal are not all the same. Variation is common and is to be expected.

FREQUENCY HISTOGRAMS

The *frequency histogram* is one tool that helps us keep track of variation. As we mentioned in Module 1, a frequency histogram is a "snapshot" of a process that shows (1) the spread of measurements and (2) how many of each measurement there are. Figure 4-1 illustrates these points.

In Figure 4-1, which is a frequency histogram, the lower edge of the chart, called the horizontal scale, records the sizes of men's sport shoes sold during a week's time. Notice that the sizes are listed in order from left to right. In this example, each number really represents a group, or class: 6 represents sizes 6 and 6½; 7 represents sizes 7 and 7½; and so on.

The vertical scale along the left-hand edge records how often each size sold. As you can see, size 6 sold 10 times; size 7 sold 16 times; size 8 sold 24 times; and so on.

Figure 4-1 tells us a great deal about variation. The sizes vary from 6 for the smallest size to 14 for the largest. The most frequent size sold is in the size 10 group. The average size sold is about 10½. This frequency

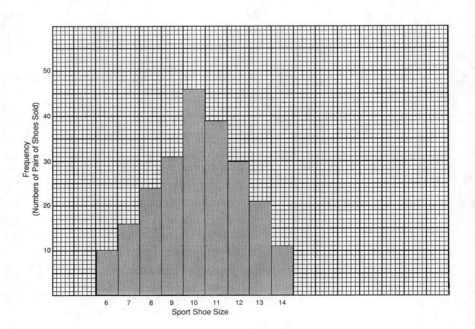

Figure 4-1. Frequency histogram of sales of men's sport shoes.

histogram shows you all these things quickly and easily without formulas or tables.

Still, frequency histograms don't tell you everything about variation. The histogram in Figure 4-1, for example, does not tell you whether the variations came from just one store or from more than one. Also, it does not show you any patterns over time. That is, you can't tell from Figure 4-1 whether the first pair of size 6 was sold on Monday or on another day. (See Module 5 for a chart that does show patterns through time.)

Before we discuss what frequency histograms can or cannot do, we will see how to construct one.

CONSTRUCTING A FREQUENCY HISTOGRAM

In this section, we will go through the steps for constructing a frequency histogram. At each step, we will describe the step and give an example.

Step 1. Collect data.

If you're lucky, you can find data already collected in reports, data files, and elsewhere. If not, you may have to collect them yourself. Divide the data into fairly small groups to make them easy to work with.

For example, a bank is studying processing times for accounts of medium sized businesses in order to distribute the workload equitably. A bank employee has recorded the times for 50 such transactions. See Table 4-1.

TABLE 4-1
Minutes to process transactions.

84	54	50	58	34
26	72	58	74	56
22	72	16	36	24
10	70	36	70	72
52	36	44	42	48
54	62	56	60	58
42	50	42	68	80
76	44	20	46	58
64	86	50	60	46
32	60	56	48	34

As you recall from Module 1, variation in a process can be measured. Table 4-1 shows the variation in minutes for the processing of these accounts. Think of the five columns in this table as five groups of data. There are only 10 numbers in each group, so the groups will be easy to work on.

Step 2. Find and mark the largest and smallest numbers in each group.

Circle the largest numbers and draw boxes around the smallest. Check your work.

In Table 4-2, we circled the largest number in each column and drew a box around the smallest. Then we checked each column.

TABLE 4-2
Minutes to process transactions. Largest and smallest numbers in each column have been marked.

(84)	54	50	58	34
26	72	(58)	(74)	56
22	72	[16]	[36]	[24]
[10]	70	36	70	72
52	[36]	44	42	48
54	62	56	60	58
42	50	42	68	(80)
76	44	20	46	58
64	(86)	50	60	46
32	60	56	48	34

Step 3. Find the largest and the smallest numbers in the whole set.

Double circle the very largest and draw a double box around the very smallest. Check your work.

In Step 2, you worked on small groups of measurements, and it was easy to find the largest and the smallest in each group. Now, in Step 3, look only at the numbers with circles and boxes around them. In Table 4-3, 86 is the largest of all the circled numbers. Put a second circle around it. The 10 is the very smallest of the numbers in boxes. Draw a second box around it. Now check to make sure you did it correctly.

TABLE 4-3
Minutes to process transactions.
Largest and smallest numbers in the set have been marked.

(84)	54	50	58	34
26	72	(58)	(74)	56
22	72	[16]	[36]	[24]
[10]	70	36	70	72
52	[36]	44	42	48
54	62	56	60	58
42	50	42	68	(80)
76	44	20	46	58
64	((86))	50	60	46
32	60	56	48	34

You may feel that this procedure takes you through too much detail, but keep two things in mind. First, we are writing this book with the expectation that you are learning this technique for the first time. Once you have learned and practiced it, you will be able to zip through the steps. Second, we are giving you details that will make it easy to use the technique and prevent you from making mistakes. Breaking this set of times into small groups makes it simple to find the largest and the smallest times in each group. Then it is easy to locate the very largest and the very smallest. Because you broke the times down into small groups, you can check your work easily and reduce the chance of making errors.

Step 4. Calculate the range of the data.

Subtract the very smallest number from the very largest number. The very largest number is 86 and the very smallest is 10, so the range is 76.

Very largest minus very smallest equals range.
$$86 - 10 = 76$$

Step 5. Determine the intervals (also known as class intervals) for your frequency histogram.

From previous steps, you know that the measurements cover an *interval* from 10 to 86. Now divide this large interval into a number of smaller intervals of equal width. One rule of thumb is to use about 10 intervals, but this number of intervals doesn't always work. See Table 4-4 for guidelines.

TABLE 4-4
Guidelines for determining the number of intervals.

NUMBER OF READINGS	NUMBER OF INTERVALS
Fewer than 50	5 to 7
50 to 100	6 to 10
101 to 150	7 to 12
more than 150	10 to 12

It is important to choose the right number of intervals for the number of readings. Too few intervals sometimes hide valuable information. Too many intervals may give such a flat histogram that you miss something important. You need to be skillful in picking the right number of intervals so that the information in the data will show up in the histogram. This skill comes with practice.

Let's try 8 class intervals for our data because Table 4-4 recommends 6 to 10 intervals for 50 readings.

Step 6. Determine intervals, boundaries, and midpoints.

First, divide the range of the data by the desired number of intervals. Round off this result for convenience. This gives the width of each interval.

The range of the set of 50 observations is 76. When you divide this range by 8 (the desired number of intervals), the result is 9.5.

$$76 \div 8 = 9.5$$

If you round off 9.5 to 10.0, which will be much easier to work with, you can group your data into eight intervals, each 10.0 units wide.

Next, set up *boundaries* for the intervals. Every reading must fall between two boundaries, for reasons we will discuss in what follows.

Since the smallest reading is 10, you may want to make the first interval go from a lower endpoint of 10 to an upper endpoint of 20, the second from 20 to 30, and so on, because you decided to make the intervals 10.0 units wide. But if you have a measurement of exactly 20, you will have the problem of deciding whether to put the 20 into the first interval (10 to 20) or the second (20 to 30).

Boundaries solve this problem. Set up boundaries between the intervals and make the boundaries such that no readings can fall on them. The easy way to do this is to add or subtract one decimal place from an endpoint. Since the data in Table 4-3 have no decimal places, subtract 0.5 from the endpoint of each interval. This changes the 10 endpoint to 9.5, the 20 endpoint to 19.5, and so on. In this case, you have subtracted, but you could just as easily have added 0.5 to the endpoints of the intervals.

Now no observation can fall on the boundaries, and the problem is solved. The first interval runs from 9.5 to 19.5, the second from 19.5 to 29.5, and so on. Table 4-5 has eight intervals, each 10.0 units in width.

Finally, set a *midpoint* at the center of each interval. (A little rounding off is all right.) The first interval runs from 9.5 to 19.5, a width of 10.0 units. Half this width is 5. Add 5 to the lower boundary, 9.5. The result is 14.5, the midpoint of the first interval. Round up to 15.0. Set all the other midpoints in the same way to obtain midpoints of 25, 35, and so on, as shown in Table 4-5.

TABLE 4-5
Midpoints, intervals, and boundaries for minutes to process transactions.

MIDPOINT	INTERVAL	BOUNDARIES
15	10–20	9.5–19.5
25	20–30	19.5–29.5
35	30–40	29.5–39.5
45	40–50	39.5–49.5
55	50–60	49.5–59.5
65	60–70	59.5–69.5
75	70–80	69.5–79.5
85	80–90	79.5–89.5

Step 7. Determine the frequencies.

Tally the data in each class interval and check the tallies. Now add them, and list the totals under "Frequency." As a final check, add all the numbers in the "Frequency" column. This final total should equal the total number of readings.

Using the setup in Table 4-5, read a number from Table 4-3 and make a tally mark beside the interval where that number fits. The first number in the bottom left corner from Table 4-3 is 32, so make a tally mark beside the interval that is defined by the "29.5–39.5" boundaries. See Table 4-6.

When you look at the completed Table 4-6, you will see that there are two tally marks in the first interval of 9.5–19.5. In the third interval, 29.5–39.5, we made four tally marks as / / / / and drew a fifth horizontally through them, ⨫. Later we added one more tally mark to make a total of 6 for this interval: ⨫ / . This method makes it easy to count the tallies, and reduces errors.

Once all the tallies are done, check by doing them again. Then total the tallies for each interval under the "Frequency" column.

You can check your work in two ways. First, count the tally and the tally check to make sure each gives the same result. Then add up the entries in the "Frequency" column. The sum is 50, which you already know is the total number of readings given in Table 4-3.

TABLE 4-6
Tally and frequency of readings in each interval.

MIDPOINT	INTERVAL	BOUNDARIES	TALLY	TALLY CHECK	FREQUENCY
15	10–20	9.5–19.5	//	//	2
25	20–30	19.5–29.5	////	////	4
35	30–40	29.5–39.5	⨫ /	⨫ /	6
45	40–50	39.5–49.5	⨫ ////	⨫ ////	9
55	50–60	49.5–59.5	⨫ ⨫ ///	⨫ ⨫ ///	13
65	60–70	59.5–69.5	⨫ /	⨫ /	6
75	70–80	69.5–79.5	⨫ //	⨫ //	7
85	80–90	79.5–89.5	///	///	3
					50

Step 8. Prepare the frequency histogram.

There are two main principles to follow in preparing the frequency histogram. It should:

- tell the story of the data, no more and no less
- be neat and easy to read

In drawing the frequency histogram you must:

- mark and label the vertical scale
- mark and label the horizontal scale
- draw in the bars according to the tallies
- label the histogram

Figure 4-2 shows the tallies in the form of a frequency histogram.

The vertical scale is labeled "Frequency" and the horizontal scale is labeled "Times to Process Transactions." The intervals are identified by their midpoints. (The 10–20 interval is labeled 15, and so on.) The tally for the 15 midpoint is two units high, the tally for the 25 midpoint is four units high, and so on. Each bar has a width of 10 units: the first goes from 10 to 20, the second from 20 to 30, and so forth. Finally, the label in the upper left-hand corner, "Frequency Histogram of Minutes to Process Transactions," identifies the histogram.

Does the histogram tell the story of the data? Let's see. First, there are eight bars in the histogram. Table 4-4 recommended 6 to 10 intervals for a set of 50 readings, so 8 intervals are O.K. As far as we know, there is nothing unusual about these data that doesn't show up on the histogram, so we think this histogram tells the story of the data.

Is this histogram neat and easy to read? You can gain eye appeal in several ways, as we did here: paste graph paper onto a white background

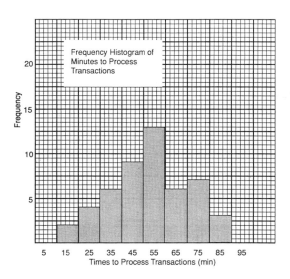

Figure 4-2. Frequency histogram of minutes to process transactions.

and write on the background; make the frequencies and midpoints easy to read; don't make the histogram too tall, too short, too wide, or too narrow. Above all, keep the histogram simple. Don't include extra information or superimpose another histogram on this one.

SOME CAUTIONS

In preparing a frequency histogram, you must be careful about several things so that the histogram will tell the story of the data it represents. The following guidelines will help you to accomplish this.

1. *Use equal-width intervals.* Unequal-width intervals tend to be confusing. Figure 4-3 is poor because the reader may miss the fact that the shipping distances do *not* go up by 10 units from one interval to the next. Figure 4-4 is just as bad. Which interval has more readings, the 35-midpoint interval or the 65-midpoint interval? Which is bigger, the 10- or the 95-midpoint interval? The reader may not get the information that was intended.

Figure 4-3. Unequal-width intervals.

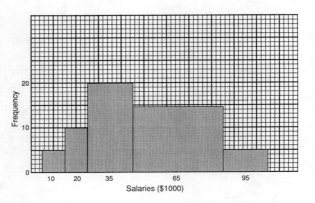

Figure 4-4. Unequal-width intervals.

2. *Do not use open intervals.* That is, make sure every interval has definite boundaries. Figure 4-5 is an example of a histogram with open intervals. What is the proper midpoint of the 40+ interval? How large is the largest number in the 40+ interval? It could be in the millions, for all the histogram tells us.

Figure 4-5. Open interval.

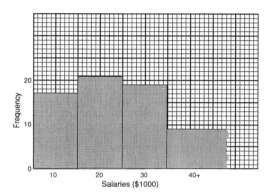

3. *Do not make any breaks in the vertical or horizontal scales.* If you do, they may be overlooked. In Figure 4-6, did you see that the first interval is seven times as tall as the second one? And did you notice that the two intervals on the end, 90 and 100, are far away from the others?

Figure 4-6. Breaks in horizontal and vertical scales.

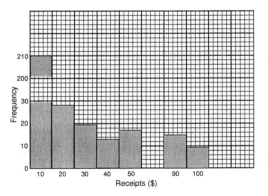

4. *Do not have too few or too many intervals.* Figure 4-7 shows data for gasoline miles per gallon, but the figure has only two intervals. By using so few intervals, the histogram hides the fact that there is one high reading of 25.21 mpg, which is very different from all the others. The first interval runs from 7.00 to 16.99, but there's no way to tell from this histogram that most of the readings are at one end of the interval.

Figure 4-7. Too few intervals.

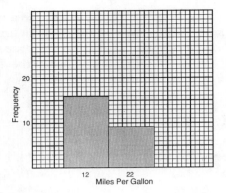

Figure 4-8, for bank account transaction times, shows the other extreme. There are so many intervals in this histogram that the horizontal axis goes on and on. It is so flat that it is hard to see the true pattern of the frequencies. We will talk more about these patterns a little later.

Figure 4-8. Too many intervals.

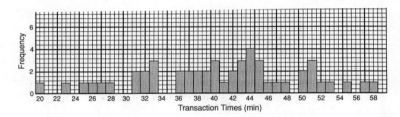

Figure 4-9. Too much information.

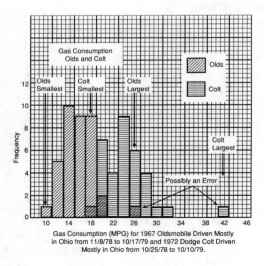

5. *Do not put too much information on one histogram.* It can be confusing. Figure 4-9 combines two frequency histograms in one chart. One shows

miles per gallon for an Oldsmobile and the other for a Dodge Colt. This figure is a mess. It may be interesting to study it carefully to see what is there, but most people would look at it quickly, blink, and go on, hoping to find something easier to understand. Since the two histograms overlap and we have used two cross-hatching methods, one for each histogram, it is hard to see where one ends and the other begins. We might be able to show the two histograms by using two different colors, but not in black and white, as is done here.

6. *Give everything needed to identify all the information completely and to make the graph understandable.* Figure 4-10 suffers from a lack of necessary information. What does the vertical scale represent? Dollars? Something else? Does the horizontal scale show gallons, bales of hay, or something else? What is a "Colt"? An automobile? A young horse? While you don't want too much information on a histogram, this figure goes to the other extreme.

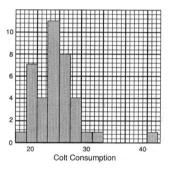

Figure 4-10. Too little information.

WHAT FREQUENCY HISTOGRAMS TELL YOU ABOUT UNDERLYING FREQUENCY DISTRIBUTIONS

We learned at the beginning of this module that there are variations between individual units, such as rifle target shots, temperatures at a resort, and times to order and pick up a meal at the fast food drive-through window. We learned that a good way to describe these variations is to build a frequency histogram.

The frequency histograms that you develop on your job will usually be based on samples. Even though they may come from the same process, your histograms will look different because the samples are different.

Suppose we fill a bucket with 1000 small metal disks coated with plastic and stir them thoroughly. We draw out a sample of 10 disks and measure the thickness of the coating on each one. We put the sample back, stir the bucket, and take another sample of 10. The frequency histograms for these two samples are shown in Figures 4-11 and 4-12.

Figure 4-11. Coating thickness, sample 1.

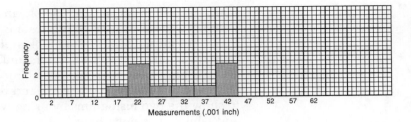

Figure 4-12. Coating thickness, sample 2.

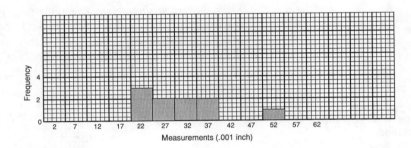

Figure 4-13. Underlying frequency distribution, coating thickness for entire bucket of disks.

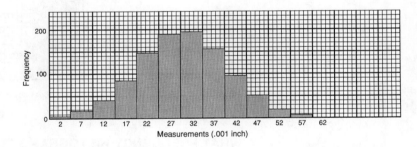

The histogram in Figure 4-13 shows what happens when we measure all 1000 disks. The pattern created by taking all 1000 disks as our sample is called the *underlying frequency distribution*. The underlying frequency distribution will always create the same pattern because it includes all the disks, which are the same every time.

By contrast, the two 10-disk samples are so small that they don't give a clear idea of the underlying distribution. In addition, the samples are so small that the two histograms are different from each other. (Compare Figures 4-11 and 4-12.) Histograms based on small samples like this will usually differ from each other because it is not likely that you will pick the same 10 disks twice.

The bigger the sample, the more the histogram will look like the underlying distribution and show what is really "in the bucket." For this reason, we recommend that you take samples of at least 50 pieces. A sample of 100 is even better.

Frequency histograms that are based on small samples will tell you something about the averages, even though they don't show the underlying frequency distributions. When you compare the histograms in Figures 4-11 and 4-12 to the underlying frequency distribution of Figure 4-13, you can see that the averages of both histograms for the smaller samples are about 30; so is the average of the underlying frequency distribution.

FREQUENCY HISTOGRAMS IN SERVICE SITUATIONS

A frequency histogram can show different situations in the process of producing or delivering a service, some good and some not so good. Figure 4-14 shows a good situation. The amount of variation is so small that all the readings on the histogram are inside the specifications. In addition, the process for delivering this service is centered at the midpoint between the specifications. What could be better!

Figure 4-14. A good situation: process spread is narrow and is centered between specifications.

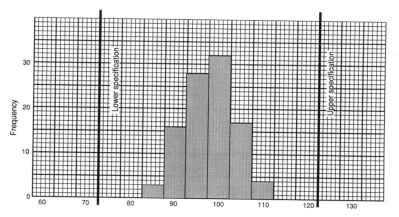

Figure 4-14 shows a good situation, but we must give you a word of warning. Even though all the readings pictured in this histogram lie inside the specifications, it is still possible that a very small percentage of the readings falls outside. You can't tell one way or the other unless you use another technique, which you will learn in Module 8.

Also, remember that a histogram is a "snapshot" of your process. It doesn't tell you anything about the process over time. To see what your process is doing over time, you need a "moving picture." In Module 5, you'll learn to construct control charts, which will give you that type of information.

Figure 4-15 shows a process in trouble. The number of days to complete projects is low in some cases, and in others it is too high. Moving the center of the process won't help. In fact, doing so will probably put a larger portion of the process outside the specifications. There's just too much variation in this process for the given specifications.

An employee probably can't solve this problem. It's up to manage-

Figure 4-15. Inherent variation is too large for specifications.

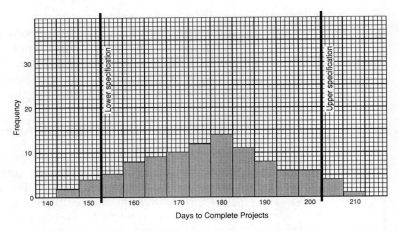

ment to decide what to do. They might change the process in some way to reduce the variation and make it look more like the process shown in Figure 4-14. They could widen the specifications. They might even decide that everybody will just have to live with the situation for the time being.

Sometimes the frequency histogram shows a process off center, as in Figure 4-16. Adjusting the process halfway between the lower and upper specifications would correct this problem. A histogram will tell you whether an adjustment is needed and how much of an adjustment to make.

Even if you can center the process perfectly, a small portion of weights may still fall outside the specifications. Even when centered, the process shown in Figure 4-16 is not as good as the process in Figure 4-14.

Figure 4-16. Process is off center.

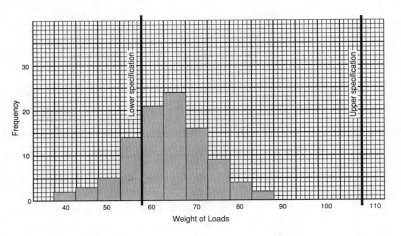

Figure 4-17 is actually two histograms. Many times, the number of servings per case of French fries is outside the specifications on both the

Figure 4-17. Two distributions present in one sample.

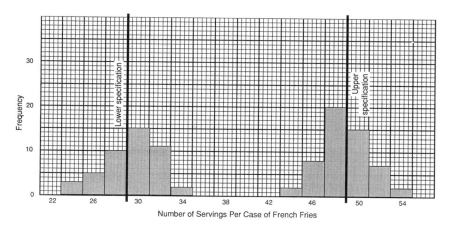

high and the low sides. This pattern suggests that there may be two underlying frequency distributions, not one. There are many possible reasons for two distributions, such as two scales weighing the fries, two suppliers to the restaurant, or two restaurant employees weighing out the servings differently.

To correct the situation shown in Figure 4-17, you must understand what is going on. You may be able to adjust one process upward and the other down, and so bring the two histograms into focus. If two scales are being used or if two shifts are doing the packaging, you need to find which histogram comes from which scale or which shift before you can make any adjustments.

There are many other patterns in frequency histograms, and each pattern will tell you something about what's going on in your process. For further information see some of the books listed in the "Recommended Readings" section.

CHECKSHEETS

As you learned in Module 3, variable data are data that come from things you can measure such as weight, time, and distance. Frequency histograms are useful for analyzing such data but sometimes we have a different kind. Then we must use other tools.

Many times the data we need to work with are not variable. We may want to know whether or not a series of tasks have been completed. Or we need to know how many of each of several categories occur, such as kinds of errors. In a continuous improvement project, the location of errors becomes important. Or we want to compare the places offering a service and the respective services being offered.

The *checksheet* is a tool that helps us collect both variable and nonvariable data and analyze them. Since it is an organized way to record information, it makes the job of collecting and analyzing data easier. A checksheet is simply a form on which we can record data in an organized manner.

There are many kinds of checksheets. We will describe five of the most common ones. We use the first type, the frequency histogram checksheet, to collect variable data. The other four help us collect non-variable data.

The Frequency Histogram Checksheet

The frequency histogram checksheet is a special type of frequency histogram. Like the histogram, this checksheet gives a great deal of information: the center of your data; the amount of variation; and the distribution of the data.

Figure 4-18 is a frequency histogram checksheet. Weights of hand-dipped ice cream cones are recorded directly onto the checksheet. By using the checksheet, data are handled only once—they are not recorded on one piece of paper and then organized and plotted on a separate histogram—each observation is written as an "X" directly on the checksheet. This means less opportunity for errors in transferring the data. It may also mean less time spent.

Use the simpler frequency histogram checksheet method when you already have some familiarity with the data. You need to know the approximate value of the smallest observation you are likely to get, as well as the largest. One problem with this checksheet is the difficulty in setting it up so that all observations will fit on it. One of the points in Figure 4-18 was outside the chart. You also need to know what the intervals are for the bottom, horizontal axis of your frequency histogram checksheet.

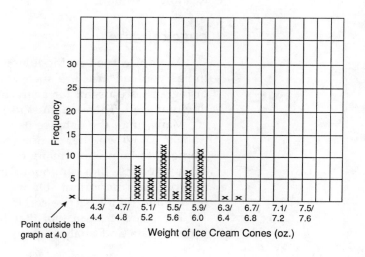

Figure 4-18. Frequency histogram checksheet showing frequencies of weights of ice cream cones. One point is outside the graph.

If you don't know these things, then you probably will have to construct a frequency histogram as described earlier in this module.

If the order in which you record your data is important so that you can look for trends through time, then you must record the data in a way that preserves the time factor of your observations. You can't plot them directly onto the histogram checksheet as in Figure 4-18. By plotting directly on the checksheet, you lose the time component, or order, of the data, as we did in Figure 4-18. If the order or time factor is not needed, then you probably can use this kind of checksheet.

The Checklist Checksheet

When you think of a checksheet, you probably think of a checklist such as an airline pilot uses. When boarding a commercial airliner, you may have noticed the pilot holding a clipboard with a checklist on it. The pilot checks off each task as it is finished. This checklist is a type of checksheet because it is a form for recording data in an organized way as the data are collected.

Figure 4-19 is a checksheet for the service person who prepares a rental car for a customer. Notice that as each task is completed, the service person simply checks off that task on the checksheet.

There are many places to use this type of checksheet in service industries. You could use one to make certain that all your materials will be

Figure 4-19. Car servicing checksheet or checklist.

Car Servicing Checksheet

Exterior:

Clean . . . . _____
Headlights O.K. . . . . _____
Parking lights O.K. . . . . _____
Signal lights O.K. . . . . _____
Tires inflated . . . . _____
Tires visibly O.K. . . . . _____
Gas tank full . . . . _____
Scratches/dents recorded . . _____

Engine compartment:

Oil level O.K. . . . . _____
Transmission fluid O.K. . . . _____
Radiator fluid O.K. . . . . _____
Belts tight . . . . _____
Air conditioner O.K. . . . . . . _____
Visual check O.K. . . . . _____

Trunk:

Clean . . . . _____
Spare tire pumped up . . . . . _____
Jack in place . . . . _____

Interior:

Clean . . . . _____
Ashtrays clean . . . . _____
All windows clean . . . . _____
All damages repaired . . . . . . _____
Panel lights work . . . . _____
Radio works . . . . _____
Air conditioner works . . . . . _____
Engine starts & runs O.K. . . _____
Transmission O.K. . . . . _____
Brakes O.K. . . . . _____
Power steering O.K. . . . . _____

available at a conference. A teacher could use it to assure that all topics are covered in a biology class. A law firm might keep a checklist of required procedures for certain types of legal work. In problem solving, you could develop a checklist of procedures to follow in identifying and solving problems.

The Item Checksheet

Another kind of checksheet is the item checksheet. In this case, we count the number of times some item occurs. We can list the item as we come to it and count how many times it keeps occurring. Or we can already have a list with the different items on it which we then count. When completed, this checksheet shows at a glance how large each category of items is and is a partial analysis of the data. We can easily take the data from this checksheet and construct a Pareto diagram. (See Module 2.)

Use the item checksheet to record the types and numbers of transactions at a bank branch, the types and numbers of patient services needed on a particular hospital ward, or the kinds and numbers of errors made in processing insurance claims. We're sure you will think of other places you can use the item checksheet.

An example is shown in Figure 4-20. As we find defects in the billing service we record them directly onto the checksheet.

Figure 4-20. Item checksheet showing kinds of errors with bills.

Type of Billing Error	Occurrences
Price error	⫽⫽⫽ //
Quantity incorrect	⫽⫽⫽ ⫽⫽⫽ //
Color incorrect	///
Wrong item	⫽⫽⫽ ///
Customer address in error	//

The Location Checksheet

With this checksheet we can indicate, and thereby see, the physical location of whatever data we're collecting. Figure 4-21 is a diagram of the human arm. Since this patient requires numerous intravenous medications (IVs), the nurses have decided to use a location checksheet to keep track of where they give the IVs. In this way, they are able to rotate the sites of the IVs and thus spare the patient the discomfort of an IV placed too often in the same place.

The location checksheet has many uses. Restaurants can use them to

Figure 4-21. Location checksheet showing drawing of human arm. X's and dates indicate IV sites.

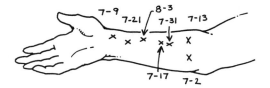

keep track of servers' table assignments. Road departments can use them to determine where potholes are occurring most often. This could help the department schedule street repairs. A life insurance agent could use a sales map as a location checksheet to see where clients are located. An office group trying to improve the layout of the department might plot traffic flow on a location checksheet to identify major traffic patterns.

The Matrix Checksheet

Figure 4-22 is an example of the matrix checksheet. In this figure, we show types of services along the top of the checksheet. Along the left-hand margin, we list six branch banks. We check off the service if it is provided by a branch bank. This arrangement is called a matrix. It is a checksheet because we record data directly on the form and they are organized.

The matrix checksheet is a powerful tool for many situations. For example, we can set up one showing kinds of errors as well as names of employees. This gives valuable information on both the kinds of errors being made and where they are happening. In solving problems and pursuing continuous improvement, the people component sometimes cannot be ignored. However, we must caution you to be very careful in using names in any analysis. We want to solve problems, *not* blame people. Be sure that your checksheet is seen as a help to your employees and not a threat. Otherwise you may win the battle while losing the war!

Figure 4-22. Matrix checksheet showing branch banks and services provided.

SERVICES PROVIDED

BRANCH BANK	ATM	Drive-Through	Teller	Lobby		
				Safe Deposit Box	Home Mortgage Loan	Business Loan
A	X	X	X		X	
B	X	X	X	X		X
C	X		X		X	X
D	X	X	X		X	
E		X	X	X		
F	X	X	X		X	X

SUMMARY

Variation is a familiar part of our lives. No two things are exactly the same. Variation is found everywhere: in bowling scores, temperatures at a favorite vacation spot, and numbers of sizes of sport shoes sold during the week.

The frequency histogram is a tool to help you deal with variation. It is a snapshot of your process, which shows the range of measurements in a sample at one point in time and tells how many measurements there are of each.

Frequency histograms have different patterns, which can reveal important information about your process. One pattern may tell you that the variation in your process is small enough so that your service can be produced or delivered within customer requirements. Another pattern may show that your process is in trouble and cannot help producing problems.

The frequency histogram can tell you about the underlying frequency distribution of the process producing or delivering your service, but it does not tell you anything about what happens over time. Other techniques, which we will discuss in later modules, will help you control and monitor your process over time.

Checksheets are tools to help you collect data. Whereas frequency histograms require variable data, such as time, weight or distance, checksheets help you handle other kinds of data.

The frequency histogram checksheet is a quick way of constructing a frequency histogram. However, you must already know approximately where the largest and smallest observations will be; you must know the intervals; and the time component of the data must not be important. The checklist checksheet allows you to check off whether each of a series of tasks is completed. By using the item checksheet, you can easily tally the items as they occur. The location checksheet gives you a way to record locations such as where tasks occur in an office. The matrix checksheet sets up a matrix on which you can record two, or more, attributes of each observation.

In continuous improvement and problem solving, we need to work on all five sources of assignable causes: equipment, method, materials, environment, and people. Although the people component cannot be left out, be careful that you do not alienate others. Don't blame employees; rather, draw them into the problem-solving process to find the causes of and solutions to problems.

PRACTICE PROBLEMS: FREQUENCY HISTOGRAMS AND CHECKSHEETS

Work through the following problems, using the statistical techniques you've learned in Module 4. The solutions can be found in the "Solutions" section beginning on page 249.

You will find that a frequency histogram worksheet form is helpful in developing your histograms. An example of such a worksheet is shown in Figure 4-23.

Figure 4-23. Frequency histogram worksheet.

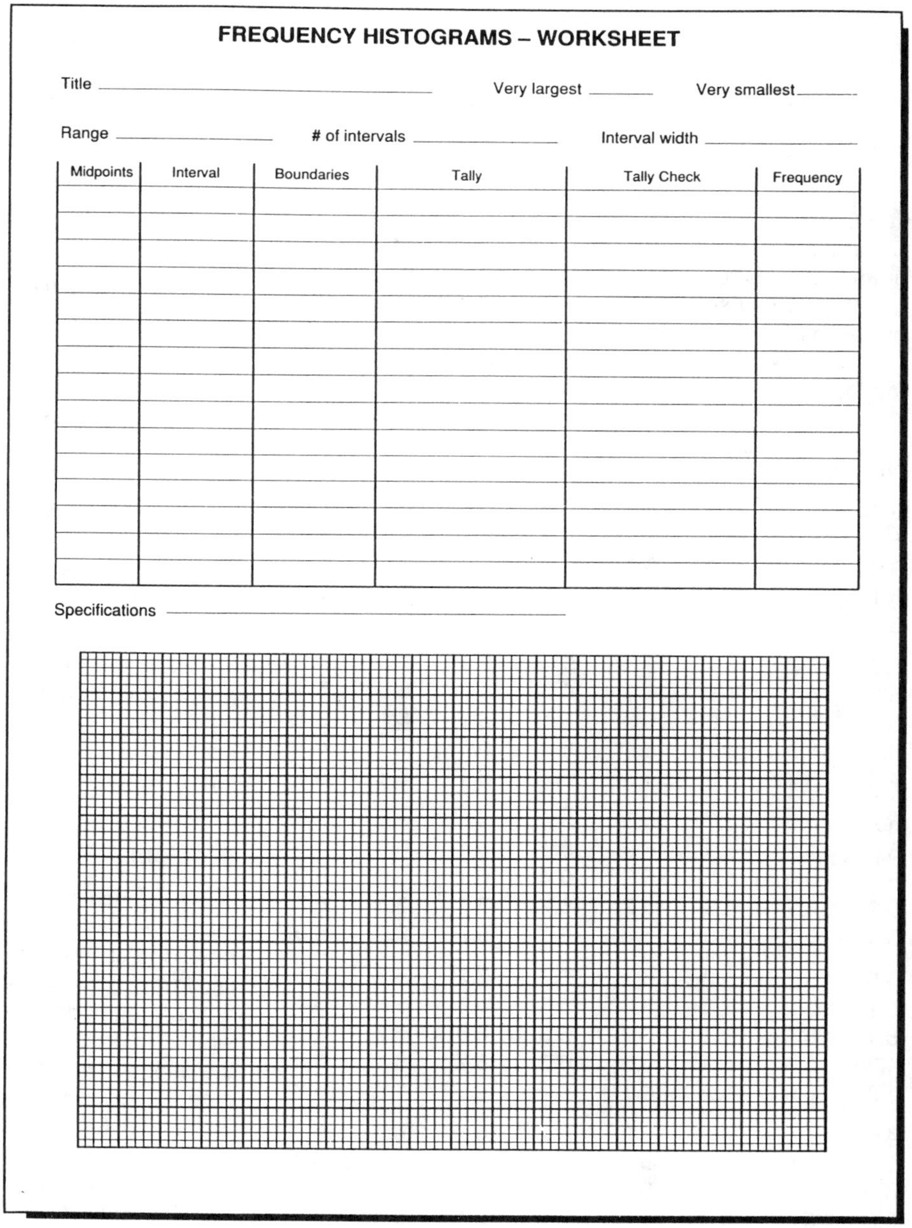

FREQUENCY HISTOGRAMS – WORKSHEET

Title _____ Very largest _____ Very smallest _____

Range _____ # of intervals _____ Interval width _____

Midpoints	Interval	Boundaries	Tally	Tally Check	Frequency

Specifications _____

Problem 4-1.

An ice cream shop is improving the training of fountain workers. Management wants ice cream cones to be consistent in appearance and weight. They are collecting data on weights of double scoops of ice cream. The following are weights of ice cream given in the order the ice cream was scooped. Reading down column A, we find the first double scoop weighed 5.3 ounces, the second weighed 5.6 ounces, the third 6.7 ounces, and so forth.

Plot frequency histograms for the data in the following columns:

1. A,B,C,D.
2. E,F,G,H.
3. I,J,K,L.
4. M,N,O,P.
5. A,E,I,M.

Compare the five histograms.
(a) Estimate the averages for the first four histograms. Do you see any differences among them?
(b) Do you see any differences among the spreads of the first four histograms?
(c) How does the fifth histogram compare to the other four? Is the spread more or less? Why or why not?

Weights of double scoops of ice cream

A	B	C	D	E	F	G	H
5.3	5.9	5.9	5.9	6.5	6.6	6.5	6.0
5.6	5.9	6.7	4.9	6.0	6.5	6.5	5.4
6.7	5.3	5.3	6.0	6.6	6.0	6.5	5.6
6.1	5.3	5.2	6.0	6.4	6.8	6.6	5.8
6.7	6.0	4.1	5.9	6.5	6.0	6.4	5.8
5.2	5.2	6.1	5.3	5.4	7.1	6.3	6.6
6.0	6.1	6.1	6.3	6.0	6.7	5.6	6.4
5.3	6.1	5.6	6.2	6.1	6.1	6.1	6.1
6.0	6.0	5.2	5.9	6.4	5.8	7.1	4.7
4.2	6.7	5.9	4.3	6.7	6.4	5.8	6.1
5.9	6.0	6.7	4.3	5.4	4.8	6.6	5.8
6.2	4.2	5.9	6.0	6.2	6.4	6.0	4.6
5.3	6.0	4.2	6.7	6.4	6.1	6.4	6.6
5.9	4.1	5.6	5.2	5.7	5.7	6.8	6.7
5.9	5.7	6.3	6.2	6.6	6.0	6.0	6.6
5.5	5.7	6.2	6.1	5.7	7.2	5.7	6.5
6.1	5.2	5.2	5.3	6.5	6.3	5.6	6.0
6.1	5.5	5.3	5.7	5.8	6.0	5.7	6.6
4.3	5.3	5.9	5.5	6.7	6.7	6.1	6.5
5.3	6.0	5.2	6.1	4.7	4.8	6.7	5.8
4.3	5.5	5.6	6.3	6.2	6.1	6.9	6.7
6.7	5.9	4.2	5.9	5.4	6.5	6.1	6.4
6.0	4.2	6.1	5.4	5.8	6.0	5.8	4.7
6.0	5.3	5.9	5.3	6.7	6.1	6.4	5.7
5.6	5.2	6.1	6.0	6.5	5.4	5.7	6.4

I	J	K	L	M	N	O	P
6.6	7.0	7.6	6.6	5.3	4.9	5.2	5.4
6.8	6.9	7.3	6.9	4.5	5.1	5.1	5.2
6.7	6.9	6.3	6.5	5.1	4.2	5.5	4.5
7.0	7.0	7.0	6.8	4.9	4.7	4.7	4.8
7.2	6.9	7.1	7.0	5.5	5.9	4.4	4.3
6.6	7.0	6.6	7.1	4.4	5.1	4.8	5.1
6.9	7.0	6.9	6.9	4.8	4.1	4.3	5.2
6.6	6.7	7.0	5.9	5.4	5.5	5.2	4.4
6.6	7.2	6.9	6.6	4.8	4.7	5.2	5.1
6.6	6.5	6.5	6.6	4.8	4.6	4.8	5.1
7.0	6.9	7.1	6.6	4.4	5.2	4.5	5.5
6.7	7.0	6.9	6.5	5.3	5.2	4.5	4.8
6.4	6.9	7.6	7.0	4.2	4.7	4.7	4.4
7.2	7.3	6.6	6.5	4.8	4.9	5.4	5.2
6.3	7.1	6.9	6.0	5.4	5.5	4.8	4.4
7.3	6.9	7.0	6.9	4.7	5.2	4.0	5.1
7.0	7.1	6.9	6.6	4.1	5.4	5.2	5.2
6.4	6.5	6.5	7.2	5.2	5.1	5.5	5.0
7.4	6.5	6.9	7.2	5.3	5.6	3.5	3.3
6.5	7.0	6.9	6.3	4.7	3.3	3.7	3.8
7.0	6.6	6.6	6.9	4.4	3.7	4.1	4.7
6.5	6.5	6.7	6.5	4.4	5.4	4.4	5.1
6.3	7.2	7.1	6.9	5.5	4.4	5.6	5.4
6.7	7.6	6.8	7.0	4.1	5.3	4.5	4.5
6.8	6.6	7.0	7.3	3.8	3.9	3.3	4.7

Problem 4-2.

A branch bank is working to improve drive-through customer service. As a first step, it decided to find out how long customers actually spend at the drive-through window. Each reading is the time from when a customer's car stopped at the window to when the customer started to drive away. Time spent waiting in line to get to the window was not recorded. The times are given in order of occurrence, that is, the first time represents when the first customer arrived at the window, the second time is for the second customer, and so on.
(a) Construct a frequency histogram of these data.
(b) About where is the average of all the data? Use the "eyeball method."
(c) Is your frequency histogram "bell-shaped?" Would you expect it to be? Why or why not?
(d) Are any of the data points suspicious in any way? Why?

Time at the Drive-Through Window (5:48 means 5 minutes, 48 seconds)

Order of Arrival	Time at Window	Order of Arrival	Time at Window	Order of Arrival	Time at Window
1	5:48	20	5:07	39	2:17
2	6:21	21	2:03	40	2:30
3	3:36	22	3:48	41	6:49
4	6:03	23	3:39	42	3:21
5	8:47	24	4:34	43	1:37
6	6:04	25	1:23	44	3:26
7	3:17	26	1:00	45	2:28
8	2:28	27	2:36	46	1:07
9	3:36	28	2:41	47	1:40
10	1:31	29	2:37	48	1:38
11	3:51	30	2:00	49	5:47
12	6:27	31	2:45	50	0:48
13	7:30	32	0:20	51	1:55
14	6:41	33	3:29	52	3:36
15	3:25	34	0:55	53	1:56
16	4:47	35	1:29	54	5:35
17	1:20	36	2:24	55	2:50
18	3:52	37	1:53	56	1:43
19	2:42	38	2:22		

Problem 4-3.

(a) Suppose you are planning to buy a personal computer (PC) for business use at home. What characteristics would you look for? Prepare a checklist checksheet for this purchase.

(b) Set up an item checksheet for defects or problems related to a PC.

(c) In an office, there are several filing cabinets in different places. Construct a location checksheet for usage of these cabinets. Or prepare a checksheet for photocopy "blips" that are caused by the photocopier itself.

(d) Prepare a frequency histogram checksheet representing miles per gallon for your automobile.

(e) Make up a matrix checksheet. Across the top of the page, write your favorite restaurants. Along the left margin, put what's important to you as a customer.

MODULE 5
Variables Control Charts

> **NEW TERMS IN MODULE 5**
> (in order of appearance)
>
> $\bar{X}$-R chart
> mean ($\bar{X}$)
> inherent variation
> overall mean ($\bar{\bar{X}}$)
> average range ($\bar{R}$)
> D_4
> decision chart or flow chart
>
> A_2
> median and range ($\tilde{X}$-R) chart
> median ($\tilde{X}$)
> median of medians ($\tilde{\tilde{X}}$)
> median of ranges ($\tilde{R}$)
> $\tilde{A}_2$
> $\tilde{D}_4$
> individual and range (X-R) chart

Why use control charts? The following story will help answer this question.

In a munitions factory during World War II, many of the operators were young women with boyfriends or family overseas on the front lines. These operators were determined to make the best product possible for their men to use in battle. At the end of the line, an operator would weigh an artillery shell for the powder content. If it was above the standard, she would yell back to the beginning of the line and they would reduce the amount of powder. If it was below, they would increase it. The correcting went on like this all day long, up or down.

Then the company put in control charts. The charts told the operator when to correct the process and when to leave it alone. The result was fewer corrections to the process and a more consistent product. The operators didn't work as hard, but they made a better product!

We will begin this module by looking at one of the best known and most widely used control charts, the average and range, or $\bar{X}$-R ("X bar, R"), control chart. The average and range chart will help you to see whether or not you have done your work correctly.

First, you will learn how to use average and range charts that are already set up. That is, someone else has figured out the control limits, the sample size, and how often to sample. You will put the charts to work in order to tell whether or not your process is running well.

Next, you'll learn how to interpret average and range charts. What are the charts telling you if a range (R) or an average ($\bar{X}$) is close to the control chart limits? What if one point falls just outside the limits or several points in a row fall on one side of the center line?

Finally, you will have an opportunity to set up average and range charts. This is detailed work, with more arithmetic than you've used so far, but some people like to set up charts because new charts can reveal surprises about the process.

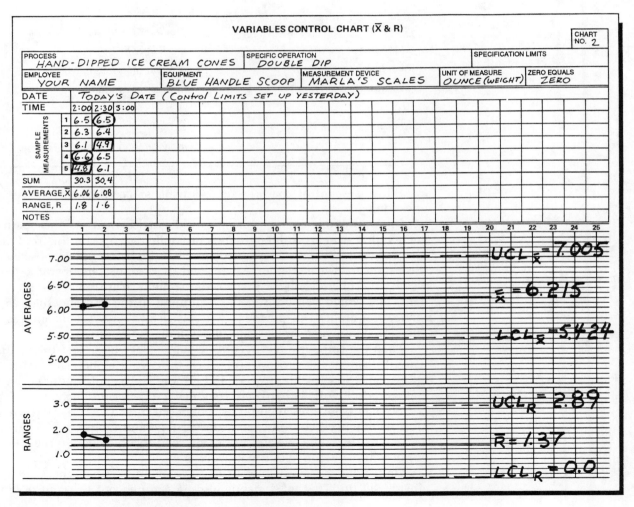

Figure 5-1. Average and range ($\bar{X}$-R) chart with two samples.

Module 5: Variables Control Charts

USING AVERAGE AND RANGE CHARTS THAT ARE ALREADY SET UP

Suppose you are a fountain worker at an ice cream shop. Two scoops of ice cream in a double-dip cone are supposed to weigh 6 ounces. You have been asked to weigh five of these cones every half hour. Someone else, maybe a supervisor, set up the control charts yesterday. The supervisor has done the first two samples for today. Figure 5-1 shows the control chart.

It's 3:00 and the supervisor is off on a break. Now it's your turn. Weigh the next five ice cream cones you scoop. The first weighs 6.9 ounces, so write 6.9 on the chart just under "3:00." The next reads 6.5 ounces, so write 6.5 on the chart. Continue in this way for the other three cones, which weigh 6.9, 7.3, and 5.9. If anything seems out of the ordinary, write it down in the "NOTES" row under "3:00." See Figure 5-2.

Figure 5-2. Portion of X̄-R chart with new measurements.

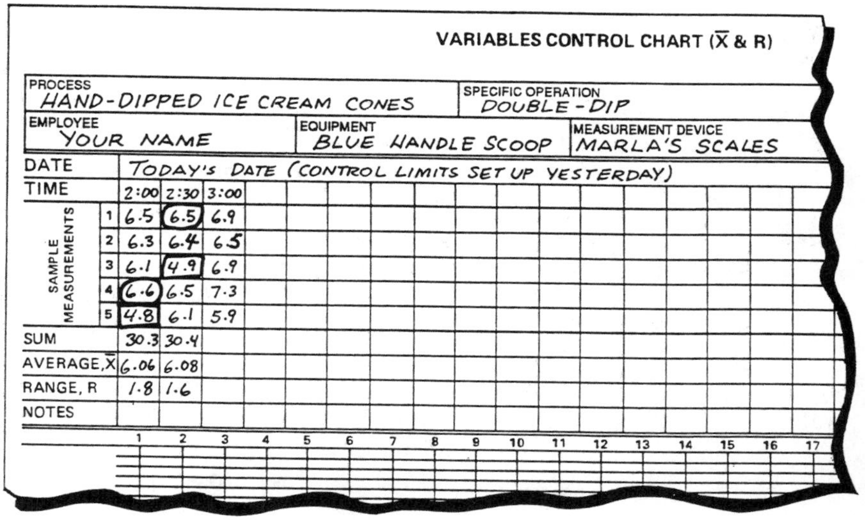

		VARIABLES CONTROL CHART (X̄ & R)		
PROCESS	HAND-DIPPED ICE CREAM CONES		SPECIFIC OPERATION DOUBLE-DIP	
EMPLOYEE YOUR NAME		EQUIPMENT BLUE HANDLE SCOOP		MEASUREMENT DEVICE MARLA'S SCALES
DATE	TODAY'S DATE (CONTROL LIMITS SET UP YESTERDAY)			
TIME	2:00	2:30	3:00	
SAMPLE MEASUREMENTS 1	6.5	6.5	6.9	
2	6.3	6.4	6.5	
3	6.1	4.9	6.9	
4	6.6	6.5	7.3	
5	4.8	6.1	5.9	
SUM	30.3	30.4		
AVERAGE, X̄	6.06	6.08		
RANGE, R	1.8	1.6		
NOTES				

Now add up the five measurements and write the result, 33.5, in the "3:00" column in the "SUM" row. See Figure 5-3. Next, figure out the average, which is sometimes called the *mean*, or X̄. (You have already seen the symbol X̄ in Module 1.) The average is simply the total, or sum, of the weights divided by the number of weights.

Average equals total divided by number of weights.
$$\bar{X} = 33.5 \div 5 = 6.70$$

Since your sum was 33.5 and you weighed five cones, simply divide 33.5 by 5. A calculator makes it easy, or if you're good with figures, do it in your head. (A trick that works only with *five* numbers is (a) multiply

Figure 5-3. Portion of X̄-R chart with sum, average, and range.

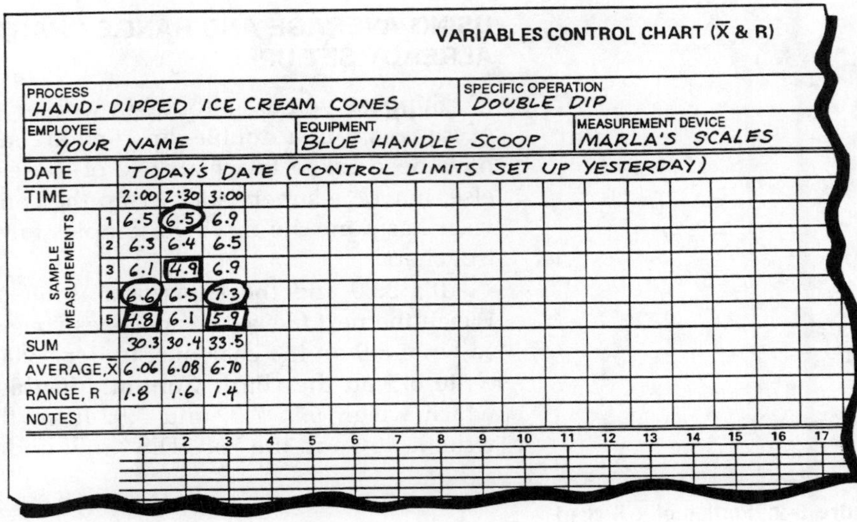

			VARIABLES CONTROL CHART (X̄ & R)		
PROCESS HAND-DIPPED ICE CREAM CONES			SPECIFIC OPERATION DOUBLE DIP		
EMPLOYEE YOUR NAME			EQUIPMENT BLUE HANDLE SCOOP	MEASUREMENT DEVICE MARLA'S SCALES	
DATE	TODAY'S DATE (CONTROL LIMITS SET UP YESTERDAY)				
TIME	2:00	2:30	3:00		
SAMPLE MEASUREMENTS 1	6.5	6.5	6.9		
2	6.3	6.4	6.5		
3	6.1	4.9	6.9		
4	6.6	6.5	7.3		
5	4.8	6.1	5.9		
SUM	30.3	30.4	33.5		
AVERAGE, X̄	6.06	6.08	6.70		
RANGE, R	1.8	1.6	1.4		
NOTES					

the sum, 33.5, by 2 for a total of 67.0 and (b) move the decimal point *one* place to the left.) The result should be 6.70, which you enter in the "AVERAGE, X̄" row. See Figure 5-3.

Finally, figure the range of the sample data. Find and mark with a circle the largest weight in your 3:00 sample, 7.3. Draw a box around the smallest, 5.9. Find the range as follows:

Range equals largest weight minus smallest weight.
$$R = 7.3 - 5.9$$
$$R = 1.4$$

Write 1.4 on your chart in the "RANGE, R" row. The top part of your chart looks like Figure 5-3.

Next plot the range and the average as small dots on the graph and draw a line to connect each new dot with the previous dot. Your chart should look like Figure 5-4.

As you learned in Module 1, averages and ranges have control limits marked by UCL (upper control limit) and LCL (lower control limit). Keep that in mind as you look at what you plotted. First, is the range where it's supposed to be? The range is 1.4, and that falls between the lower control limit of zero and the upper control limit of 2.89, so the range is O.K. Second, is the average where it is supposed to be? Yes—it's 6.70, which falls between the lower control limit of 5.424 and the upper control limit of 7.005.

What do these plots tell you? When the range *and* the average are inside the control limits, your process is stable. It's "in statistical control." Don't make any adjustments—just keep scooping.

Figure 5-4. Portion of X̄-R chart with new average and range plotted.

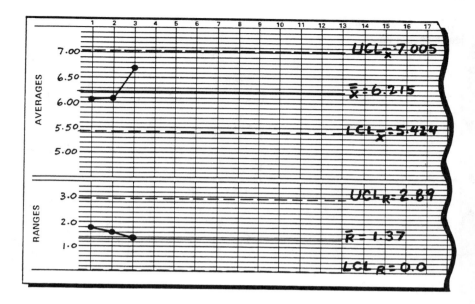

INTERPRETING AVERAGE AND RANGE CHARTS

Now that you know how to figure averages and ranges and how to plot the points, take a few more samples and plot their ranges and averages. Then you can interpret average and range charts.

On your control chart (Figure 5-5), every point is inside the control limits. This tells you to keep running the process. No changes or corrections are required because all points are inside the limits.

You may ask, "What about the fifth average, 6.88, which is very close to the upper control limit of 7.005?" The rule is clear-cut: If the point is inside, make no corrections and keep on producing. If the point is outside, find the assignable cause and correct it.

Why do the points move up and down inside the limits? What you see in Figure 5-5 is most likely natural variation in the process. This is called *inherent variation* and is due to chance causes. Usually, the employee cannot get rid of it because it is a part of the process itself. For example, if you're waiting on customers, there's variation in the number of people waiting to be served. Then, too, any time there is a manual operation such as scooping ice cream, there will be variation in the process. All these variations are natural, and all are beyond the operator's control. But these inherent variations pull the ranges and averages up and down *within* the control limits. (This is true *only* for inherent variation. We are not including variations due to assignable causes, such as a new tub of ice cream that is too hard to scoop.)

Figure 5-5. All points are inside control limits.

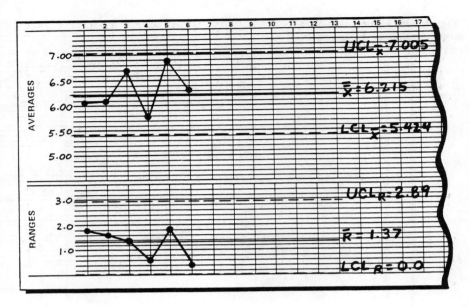

AVERAGES OUTSIDE CONTROL LIMITS

When an average is outside the control limits, as in Figure 5-6, what does the chart tell you to do? It tells you to adjust the process, to correct it. For some reason, the process has shifted up and needs to be corrected downward. How do you make the correction? There is no simple answer. In the case of weights of ice cream cones, you might have to adjust your method of scooping. For another process, say photoduplicating reports, you might have to clean the glass to remove dust and lint so that you don't get blips on your copies. Some corrections are easy to make, and some are not.

How do we interpret the average outside the limits in Figure 5-6? Because it is outside, the process is now out of statistical control. This one average shows that simple inherent variation is no longer at work. Inherent variation is now combined with something more—an assignable cause. The assignable cause changes the overall average of the process. Once that happens, the averages begin to fall outside the control limits. That is why our seventh average, 7.24, is outside the limits.

Usually, an assignable cause points to a problem that you can handle yourself in your immediate work situation. At other times, you will have to notify your boss because of the nature of the problem, such as a change in material used to produce or deliver your service. We are not talking here about who is responsible. Instead, we want you to see that *something, somewhere*, has changed. The point outside the control limits tells you just that.

Figure 5-6. Average outside control limits.

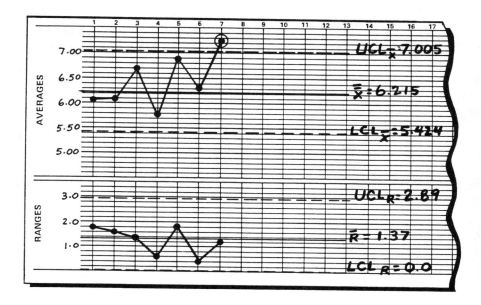

What's special or assignable about this point outside the limits? It's not natural to the process, it doesn't happen all the time, and, most importantly, you may be able to identify the cause—that's why it's called an assignable cause. In the case of the weights, your sample average may have gone outside the limits just when you started to dip from a new tub of ice cream. Careful checking might show that the new tub was colder and harder than usual. In this case, weighing a few scoops could identify that there is an assignable cause. Note that the control chart tells only *when* the assignable cause occurred, not *why*.

OTHER SIGNS OF A PROCESS OUT OF CONTROL

From time to time, the points you plot may form a pattern. One particular pattern to look for is seven points in a row above or below the line marked $\bar{\bar{X}}$ in the "AVERAGES" section or the line marked $\bar{R}$ in the "RANGES" section. What is such a pattern telling you? This is another signal—a weaker one—that the process is out of control, even though all the points are still inside the control limits.

There are other patterns of longer runs of points, as well as too many or too few points close to the process average. To interpret these patterns, we suggest that you look them up in Ishikawa's *Guide to Quality Control*, in Ott's *Process Quality Control*, or in Grant and Leavenworth's *Statistical Quality Control*. If you don't own any of these books look for them at your public library.

SOURCES OF ASSIGNABLE CAUSES

The sources of assignable causes often come from one of several main categories, which you can find on the fishbone diagram in Module 1 (see Figure 1-4) and which we discussed in Module 2. These include the following:

(1) *Equipment or machine.* By equipment, we mean that the piece of equipment itself has changed somehow, and that's why the sample average went outside the limits.

(2) *Materials.* The assignable cause category of materials means that something has changed in the materials themselves. This change has caused the average to go outside the control limits.

(3) *Method* is the way we do things. Switching from one method to another may cause the average to go outside the control limits.

(4) *Environment* may also be a source of variation.

(5) You, the *employee.* We listed you last because in our opinion, it's usually not the employee that is the cause. Most of the time it turns out to be one of the other four. Too often, people have been blamed when equipment, method, materials, or environment are the real source of the problem. But when it is a person who is the assignable cause, it may be because he or she is not adequately trained or a relief person has just come on.

RANGES OUTSIDE CONTROL LIMITS

In Figure 5-7, range number 8 is out of control. In this situation, it is helpful to ask first, "Is the inherent variation under the employee's control?" In our opinion, an employee does not usually have control over the inherent variation of the process. Therefore, a range outside the limits probably means that the process itself has gone haywire. An example of this is the temperature of the storage freezer, which would affect the softness or hardness of the ice cream. In those cases, the employee would probably have no control over the inherent variation, but the control chart shows that there is a problem. As the employee, your job is to notify management so that they can find out what has caused the change in the inherent variation and so make corrections.

One word of caution—check your arithmetic and (if possible) your measurements to be sure you didn't make a mistake in arithmetic or in measuring. Such a mistake can easily send the range outside the control limits, so avoid possible embarrassment by checking it out. *You* could be an assignable cause!

In rare cases, though, the variation is due to something you *can* control, at least partially. If this is so, you may be able to make a correction that will bring the ranges back into statistical control. These cases might include how often you defrost the freezer and how quickly the ice cream

Figure 5-7. Range outside control limits.

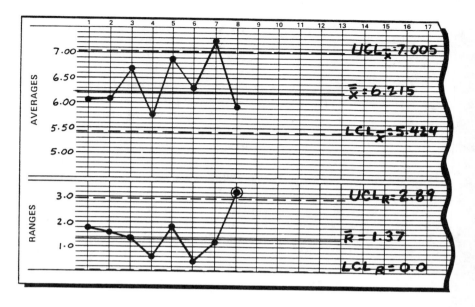

tubs are moved from the freezer to the fountain area. Ranges may go out of control because of momentary lack of attention, fatigue, or a change in method.

SETTING UP AVERAGE AND RANGE CHARTS

Now that you know how to use average and range charts, you can try setting up your own. Control charts serve several purposes. (1) They can be used for control. That is, the charts can be used to tell you whether to continue the operation or to find and correct assignable causes. (2) Charts can be used for analysis. You might be looking for the amount of inherent variation in the process of producing or delivering a service. Or you might want to find the differences between days, materials, or even employee techniques. (3) Charts can also be used for education, communication, or documentation. They may help you, as an employee, to focus attention on consistent quality.

In the case of weights of the double-scoop cone, the manager of the shop and the fountain employees have agreed that ongoing control is necessary in order to give a consistent product to the customer. If you are a fountain operator, average and range charts should help you in controlling your process. Even though control is the primary purpose, the charts will also provide some extras. They will help show your supervisor and the shop manager how well the process is doing. Then, at some later time, the charts could be used as documents to review the process or evaluate the effectiveness of training.

Step 1. Choose what to measure.

Many different things about the process of producing and/or delivering a service can be measured, so you must choose what you want to measure. There are two things to remember. First, select something important in the process and control that one thing. (See the criticality designation system in Module 1.) Don't make control charts for every characteristic in the process. Second, you may not be able to measure directly the characteristic that you choose. If you can't, you must find something *measurable* that will allow the important characteristic to be controlled with charts.

In the case of the double-dip cones, it is important that the scoops be consistent in size. Nothing is more disappointing than buying an ice cream cone one time, then going back, buying the same size, and having it appear smaller than the one before. The cones shouldn't be too big so that you aren't able to get the expected number of scoops from the tub. On the other hand, the scoops should not be undersized; otherwise customers won't be getting their money's worth.

What measurable characteristic will allow you to control the scoops of ice cream? Weight. This is something that can be measured and expressed in numbers, and numbers can be entered on control charts. In other words, you can control the scoops through the use of average and range charts. Because you can measure weights, you can give the customer good value as well as getting the expected number of servings from the tub.

Step 2. Take the samples.

In setting up average and range charts, you will need a series of samples. Each sample will consist of several measurements, often four or five. You will use the information from these measurements in several ways on the control chart, such as determining the average of each sample.

How you choose the sample is very important. Remember that the sample is to be taken in such a way that *only* inherent variation is seen within the sample. If there is an assignable cause, you want it to act on every item in the sample. Then the variation within the sample will be only inherent variation.

One way to do this is to choose your samples so that the individual pieces within each sample are as much alike as possible. Two things will help you do this. First, take each sample over a very short period of time. Second, take each sample (and all the samples for your initial control charts) from a single source of data. That is, take the sample from *one* set of materials, *one* piece of equipment such as a bank ATM, *one* employee, and so forth. (Sometimes there are so many different sources of data that it is not worthwhile to make separate control charts on each source. In that case, how many charts should you make? That depends on how many charts you need and how many charts your company is willing to pay for.)

In setting up control charts for weights, let's suppose that you took samples of your work every 30 minutes. The five cones in each sample were scooped and weighed one right after another, and you noted their order. This means that each sample came from a very short period of time, there was only one employee, and the cones came from only one tub. As a result, each sample of five cones will be as much alike as possible. Therefore, any differences or variations within a sample should be no more than the inherent variation.

VARIABLES CONTROL CHART ($\bar{X}$ & R)

PROCESS: HAND-DIPPED ICE CREAM CONES
SPECIFIC OPERATION: DOUBLE DIP
SPECIFICATION LIMITS:
CHART NO.:
EMPLOYEE: YOUR NAME
EQUIPMENT: BLUE HANDLE SCOOP
MEASUREMENT DEVICE: MARLA'S SCALES
UNIT OF MEASURE: OUNCE (WEIGHT)
ZERO EQUALS: ZERO
DATE: 4/12

TIME	11AM	11:30	12:00	12:30	1:00	1:30	2:00	2:30	3:00	3:30	4:00	4:30	5:00	5:30	6:00	6:30	7:00	7:30	8:00	8:30
1	5.3	4.4	4.4	4.7	4.4	4.9	5.1	5.2	5.2	3.7	5.2	4.8	4.5	5.0	4.1	5.4	5.1	5.5	5.1	4.7
2	4.5	4.8	5.3	4.1	4.4	5.1	4.1	5.2	5.4	5.4	5.1	4.3	4.5	5.2	4.4	5.2	5.2	4.8	5.2	5.1
3	5.1	5.4	4.2	5.2	5.5	4.2	5.5	4.7	5.1	4.4	5.5	5.2	4.7	5.5	5.6	4.5	4.4	5.0	5.4	
4	4.9	4.8	4.8	5.3	4.1	4.7	4.7	4.9	5.6	5.3	4.7	5.2	5.4	3.5	4.5	4.8	5.1	5.2	3.3	4.5
5	5.5	4.8	5.4	4.7	3.8	5.9	4.6	5.5	3.3	3.9	4.4	4.8	4.8	3.7	3.3	4.3	5.1	4.4	3.8	4.7

SUM
AVERAGE, $\bar{X}$
RANGE, R
NOTES

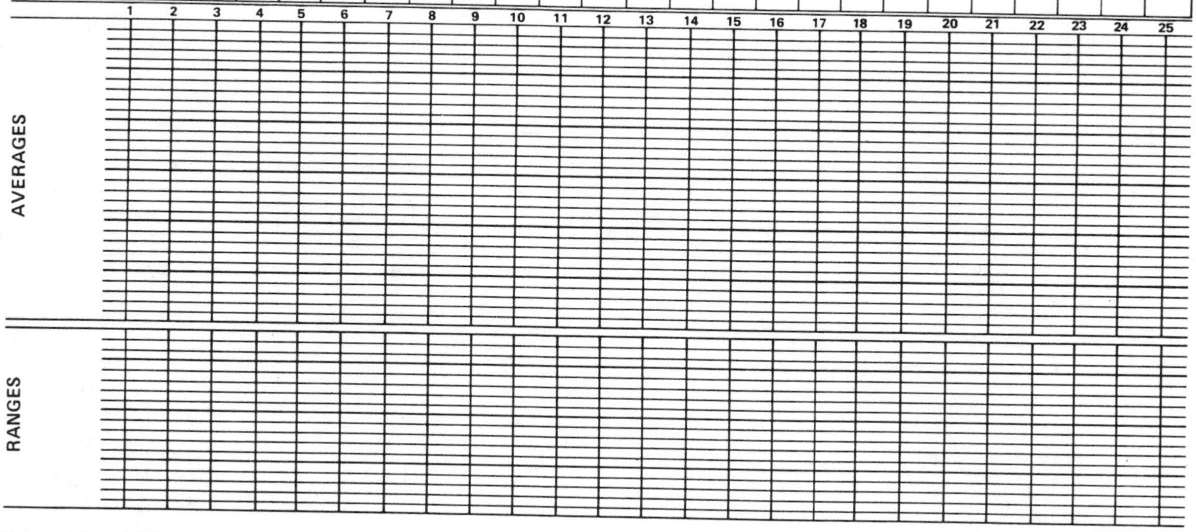

Figure 5-8. Average and range ($\bar{X}$-R) chart with background information and measurements.

Step 3. Set up forms for data and graphs.

Once you have decided on your important characteristics and your sample, it's time to think about the forms. Good forms can make the calculations easy. They have boxes for necessary background information such as date, process, service, what is being measured (hour, minute, weight), and employee.

The chart in Figure 5-8 shows boxes filled in with the type of ice cream cone, the date, and other information.

Step 4. Collect the samples and record the measurements.

Take the samples according to your plan. Weigh them in the way that you have determined, and record the weights on the form. Be sure to put the weights on the form *in the order the cones were scooped*. Record the time of each sample as appropriate. See Figure 5-8.

Step 5. Calculate the averages.

Now that you have collected the data for the initial control charts, the first thing to do is to calculate the averages ($\bar{X}$'s). For each sample, add up the weights and record the total on the form in the row marked "SUM." See Figure 5-9. Then divide this total by the number of observations in your sample and write the answer in the row marked "AVERAGE, $\bar{X}$,"

VARIABLES CONTROL CHART ($\bar{X}$ & R)

PROCESS: HAND-DIPPED ICE CREAM CONES	SPECIFIC OPERATION: DOUBLE DIP		SPECIFICATION LIMITS	CHART NO.
EMPLOYEE: YOUR NAME	EQUIPMENT: BLUE HANDLE SCOOP	MEASUREMENT DEVICE: MARLA'S SCALES	UNIT OF MEASURE: OUNCE (WEIGHT)	ZERO EQUALS: ZERO

DATE	4/12																								
TIME	11 AM	11:30	12:00	12:30	1:00	1:30	2:00	2:30	3:00	3:30	4:00	4:30	5:00	5:30	6:00	6:30	7:00	7:30	8:00	8:30					
SAMPLE MEASUREMENTS 1	5.3	4.4	4.4	4.7	4.4	4.9	5.1	5.2	5.2	3.7	6.2	4.8	4.5	5.0	4.1	5.4	5.1	5.5	5.1	4.7					
2	4.5	4.8	5.3	4.1	4.4	5.1	4.1	5.2	5.4	5.4	5.1	4.3	4.5	5.2	4.4	5.2	5.2	4.8	5.2	5.1					
3	5.1	5.4	4.2	5.2	5.5	4.2	5.5	4.7	5.1	4.4	5.5	5.2	4.7	5.5	5.6	4.5	4.4	4.4	5.0	5.4					
4	4.9	4.8	4.8	5.3	4.1	4.7	4.7	4.9	5.6	5.3	4.7	5.2	5.4	3.5	4.5	4.8	5.1	5.2	3.3	4.5					
5	5.5	4.8	5.4	4.7	3.8	5.9	4.6	5.5	3.3	3.9	4.4	4.8	4.8	3.7	3.3	4.3	5.1	4.4	3.8	4.7					
SUM	25.3	24.2	24.1	24.0	22.2	24.8	24.0	25.5	24.6	22.7	24.9	24.3	23.9	22.9	21.9	24.2	24.9	24.3	22.4	24.4					
AVERAGE, $\bar{X}$	5.06	4.84	4.82	4.80	4.44	4.96	4.80	5.10	4.92	4.54	4.98	4.86	4.78	4.58	4.38	4.84	4.98	4.86	4.48	4.88					
RANGE, R																									
NOTES																									
	1	2	3	4	5	6	7	8	9	10	11	12	13	14	15	16	17	18	19	20	21	22	23	24	25

Figure 5-9. Top part of average and range chart with sums and averages.

as you have already learned to do. Finally, check your arithmetic. (One of the best ways to check is to do it all over again.)

Your calculations for the first two samples will look like this.

<table>
<tr><td>First Sample</td><td>Second Sample</td></tr>
<tr><td>5.3</td><td>4.4</td></tr>
<tr><td>4.5</td><td>4.8</td></tr>
<tr><td>5.1</td><td>5.4</td></tr>
<tr><td>4.9</td><td>4.8</td></tr>
<tr><td>5.5</td><td>4.8</td></tr>
<tr><td>Total = 25.3</td><td>Total = 24.2</td></tr>
</table>

Average = Total ÷ number in sample
25.3 ÷ 5
= 5.06
= $\bar{X}$

Average = Total ÷ number in sample
24.2 ÷ 5
= 4.84
= $\bar{X}$

Step 6. Calculate the overall average ($\bar{\bar{X}}$).

The overall average, or *overall mean* ($\bar{\bar{X}}$), is the average of all your sample averages. First, add up all the averages ($\bar{X}$'s). Then divide this total by the *number* of averages. Check your arithmetic by repeating the calculations. (Another way to check your arithmetic is to add up all the individual readings on your chart. Count the individual weights. Then divide the total of all the weights by the number of weights.)

The weights of the double-dip cones give the following (see Figure 5-9):

Averages, $\bar{X}$'s: 5.06, 4.84, 4.82, 4.80, 4.44, 4.96, 4.80, 5.10, 4.92, 4.54, 4.98, 4.86, 4.78, 4.58, 4.38, 4.84, 4.98, 4.86, 4.48, 4.88

Total of $\bar{X}$'s: 95.9

Number of $\bar{X}$'s: 20

Overall mean, $\bar{\bar{X}}$:   95.9 ÷ 20 = 4.795

Don't enter the $\bar{\bar{X}}$ on your chart yet. Just make a note of it. You will need it later.

Step 7. Determine the ranges for the samples.

Find and circle the largest number in each sample. Find and draw a box around the smallest. Then calculate the range for each sample as follows:

Range = Largest observation minus smallest observation.
(Subtract the smallest from the largest.)

Record the ranges in the row marked "RANGE, R". Check your arithmetic.

The chart for the ice cream shows the following (see Figure 5-10):

First Sample	Second Sample
5.3	4.4
4.5	4.8
5.1	5.4
4.9	4.8
5.5	4.8
Range = 5.5 minus 4.5	Range = 5.4 minus 4.4
R = 1.0	R = 1.0

Step 8. Calculate the average range.

Add up all the ranges; then count them. Then divide the total of all the ranges by the number of ranges. The result is the *average range* ($\overline{R}$). Finally, check your arithmetic carefully. (You can't check $\overline{R}$ by calculating the range of all the data. It won't work.)

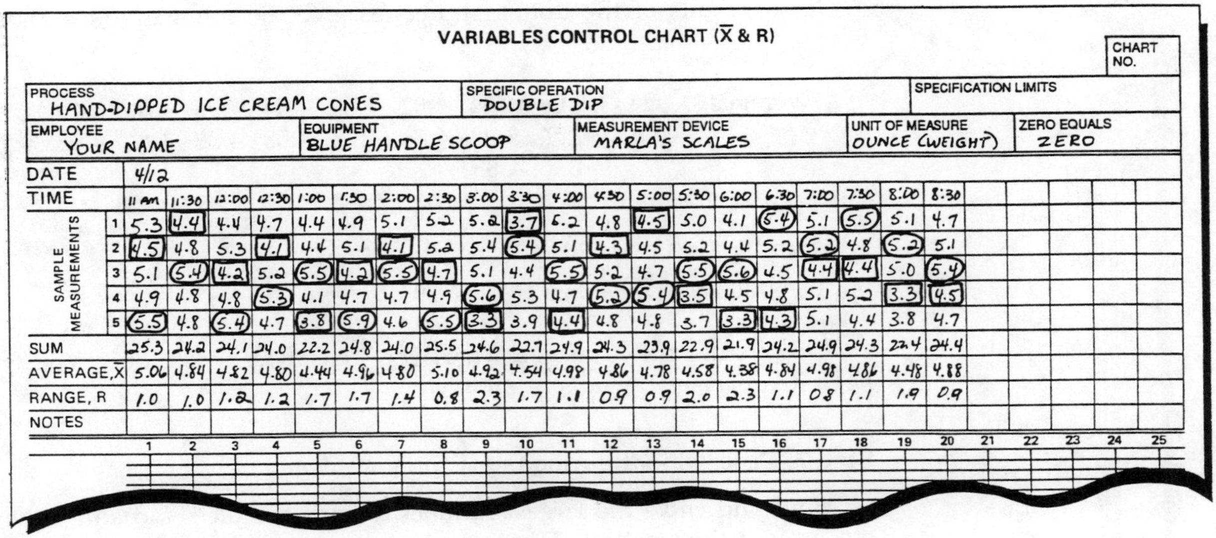

VARIABLES CONTROL CHART ($\overline{X}$ & R)

CHART NO.

PROCESS: HAND-DIPPED ICE CREAM CONES | SPECIFIC OPERATION: DOUBLE DIP | SPECIFICATION LIMITS:

EMPLOYEE: YOUR NAME | EQUIPMENT: BLUE HANDLE SCOOP | MEASUREMENT DEVICE: MARLA'S SCALES | UNIT OF MEASURE: OUNCE (WEIGHT) | ZERO EQUALS: ZERO

DATE: 4/12

	1	2	3	4	5	6	7	8	9	10	11	12	13	14	15	16	17	18	19	20
TIME	11 AM	11:30	12:00	12:30	1:00	1:30	2:00	2:30	3:00	3:30	4:00	4:30	5:00	5:30	6:00	6:30	7:00	7:30	8:00	8:30
SAMPLE MEASUREMENTS 1	5.3	4.4	4.4	4.7	4.4	4.9	5.1	5.2	5.2	3.7	6.2	4.8	4.5	5.0	4.1	5.4	5.1	5.5	5.1	4.7
2	4.5	4.8	5.3	4.1	4.4	5.1	4.1	5.2	5.4	5.4	5.1	4.3	4.5	5.2	4.4	5.2	5.2	4.8	5.2	5.1
3	5.1	5.4	4.2	5.2	5.5	4.2	5.5	4.7	5.1	4.4	5.5	5.2	4.7	5.5	5.6	4.5	4.4	4.4	5.0	5.4
4	4.9	4.8	4.8	5.3	4.1	4.7	4.7	4.9	5.6	5.3	4.7	5.2	5.4	3.5	4.5	4.8	5.1	5.2	3.3	4.5
5	5.5	4.8	5.4	4.7	3.8	5.9	4.6	5.5	3.3	3.9	4.4	4.8	4.8	3.7	3.3	4.3	5.1	4.4	3.8	4.7
SUM	25.3	24.2	24.1	24.0	22.2	24.8	24.0	25.5	24.6	22.7	24.9	24.3	23.9	22.9	21.9	24.2	24.9	24.3	22.4	24.4
AVERAGE, $\overline{X}$	5.06	4.84	4.82	4.80	4.44	4.96	4.80	5.10	4.92	4.54	4.98	4.86	4.78	4.58	4.38	4.84	4.98	4.86	4.48	4.88
RANGE, R	1.0	1.0	1.2	1.2	1.7	1.7	1.4	0.8	2.3	1.7	1.1	0.9	0.9	2.0	2.3	1.1	0.8	1.1	1.9	0.9
NOTES																				

Figure 5-10. Top part of average and range chart with ranges.

Ranges, R: 1.0, 1.0, 1.2, 1.2, 1.7, 1.7, 1.4, 0.8, 2.3, 1.7, 1.1, 0.9, 0.9, 2.0, 2.3, 1.1, 0.8, 1.1, 1.9, 0.9

Total of R's: 27.0

Number of R's: 20

Average Range, $\bar{R} = 27.0 \div 20 = 1.35$

Don't write the $\bar{R}$ on the chart yet. Save it for later.

Step 9. Determine scales for the graphs and plot the data.

First, find the largest and smallest averages ($\bar{X}$'s). Find the largest and the smallest ranges (R's). Check to make sure they are the largest and the smallest. Then set the scale for the graph in such a way that the largest and the smallest values will fit comfortably inside the ends of the scales. (The lower end of the graph for ranges is usually set at zero.) Pick scales that make it easy for you to plot the data and leave extra room on the graphs for the statistical control limits. Now plot the data and draw a line for the overall mean ($\bar{\bar{X}}$) and a line for the average range ($\bar{R}$).

Largest average ($\bar{X}$): 5.10   Largest range (R): 2.3
Smallest average ($\bar{X}$): 4.38   Smallest range (R): use 0.0

For averages, you can run your scales from 3.50 to 5.50. All the averages will fit while still leaving room for the control limits. On your form there will be five lines between 3.50 and 4.00, so each line will represent 0.1 unit. See Figure 5-11.

For the ranges, run the scale up to 3.0 at the highest heavy line. All the ranges will fit and some room will be left for control limits. Each line represents 0.2 units, which makes it easy to plot the ranges. You may be tempted to mark the three dark lines as 0.5, 1.0, and 1.5, but that arrangement won't leave much room for the upper control limit for ranges.

Finally, plot the averages and ranges as dots and connect the dots with straight lines. Draw a heavy, solid line for the overall mean ($\bar{\bar{X}}$) and mark its value at the far end. Do the same for the average range ($\bar{R}$). See Figure 5-11.

Step 10. Determine control limits for ranges.

Calculate control limits for ranges *before* you calculate the limits for averages so you will know whether the inherent variation is stable. If it is not, there is no sense in checking whether the averages are in control.

On the back of the average and range chart form is a box marked "FACTORS FOR CONTROL LIMITS." See Figure 5-12. From the table of

114 SPC Simplified for Services: Practical Tools for Continuous Quality Improvement

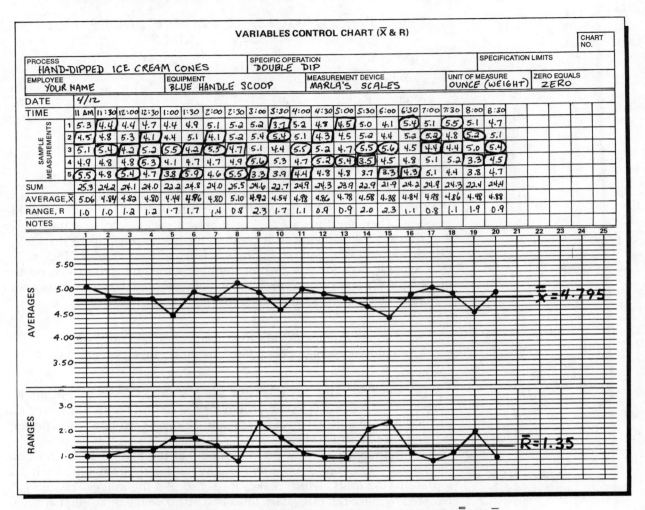

Figure 5-11. Average and range chart with data plotted on graph. Lines are drawn for $\bar{X}$ and $\bar{R}$.

average and range chart factors, choose the D_4 *factor* that corresponds to the sample size you are using. Circle the 5 under the "n" column because 5 is the size of your sample. Circle the 2.114 because that D_4 factor corresponds to a sample size of 5.

To find the upper control limit for ranges, use this formula:

Upper control limit for ranges (UCL_R) equals D_4 times $\bar{R}$.
$$UCL_R = 2.114 \times 1.35$$
$$UCL_R = 2.854$$

Figure 5-12. Factors for control limits. D₄ is circled for a sample size of five.

n	A_2	D_4	d_2	$\frac{3}{d_2}$	A_M
2	1.880	3.268	1.128	2.659	0.779
3	1.023	2.574	1.693	1.772	0.749
4	0.729	2.282	2.059	1.457	0.728
5	0.577	2.114	2.326	1.290	0.713
6	0.483	2.004	2.534	1.184	0.701

FACTORS FOR CONTROL LIMITS

Because our sample size is 5, our lower control limit will be zero.

$$LCL_R = 0.0$$

For samples of size 6 or less, the lower control limit for ranges is always zero. Be sure to check your sample size, the D_4 factor, and your arithmetic. Then draw the control limits on your range chart and label them UCL_R and LCL_R. We recommend that you draw the limits as dashed lines or colored lines so that they will be easy to see. See Figure 5-13.

Step 11. Are the ranges in statistical control?

There are three possible answers to this question: (1) *all* the ranges (R's) fall *inside* the control limits; (2) *one or two* ranges fall *outside* the limits; and (3) *three or more* ranges fall *outside* the limits.

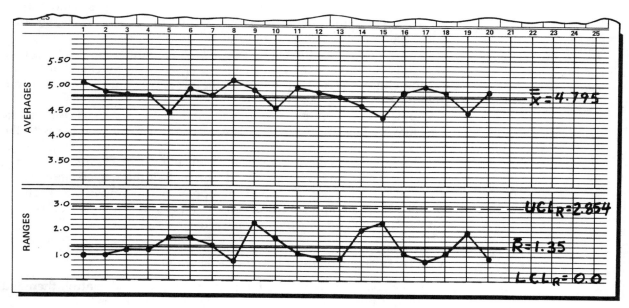

Figure 5-13. Bottom part of average and range chart with control limits for ranges.

(1) If *all* the ranges are inside the control limits—that is, if no ranges fall above the upper control limit (UCL_R) or below the lower limit (LCL_R)—the ranges are in statistical control. Now you may go ahead and figure out control limits for the averages. If a range falls exactly on the control limit, don't worry about it; it counts as inside. When we say that the ranges are in control, we are really saying that the inherent variation is stable and there are no assignable causes disturbing the ranges.

(2) Sometimes *one or two* ranges will fall outside the control limits. When this happens, it is common practice to throw out those ranges, the samples from which they came, and the sample averages. (We don't like to throw out data because even wild or extreme data can tell us something.) Then, completely refigure the overall mean ($\overline{\overline{X}}$), the average range ($\overline{R}$), and the control limits for the ranges without those out-of-control ranges.

After you refigure the upper and lower control limits for the ranges, one of two things may happen. First, one or more of the remaining ranges may still fall outside the new control limits. If that happens, the ranges are out of statistical control. In that case do *not* figure control limits for averages. Find and remove the assignable causes that are upsetting the ranges. Then set up new average and range charts with new data.

On the other hand, you may find that all the ranges are now inside the new control limits. In that case, you may go on to develop control limits for the averages. But be careful! There may still be some assignable causes that can give you trouble until you find and remove them.

(3) If *three or more* ranges are outside the original control limits, the ranges are out of statistical control and the inherent variation is not stable. Do not bother to figure out control limits for the averages. Find and remove the assignable causes that are upsetting the ranges. Then start over again—collect new data and set up new control charts.

You may use this three-step rule *only* when you are first setting up a new control chart. In the regular use of an $\overline{X}$-R control chart, one point outside either control limit tells you the process is out of statistical control.

To help you keep track of all these possibilities, a *decision chart* or *flow chart* may be helpful. Figure 5-14 is a decision chart for working with the ranges.

The ranges for the ice cream data are plotted in Figure 5-13. When you look at the range chart, you can see that all the ranges fall between the lower control limit of 0.0 and the upper control limit of 2.854. Therefore, the ranges are in statistical control. Now you can go ahead to work out the control limits for the averages chart.

Step 12. Determine control limits for averages.

Once you have found the ranges to be in statistical control, then and *only* then can you work on the control limits for averages.

On the back of the $\overline{X}$-R control chart form, look under "FACTORS

Figure 5-14. Decision chart for working with ranges.

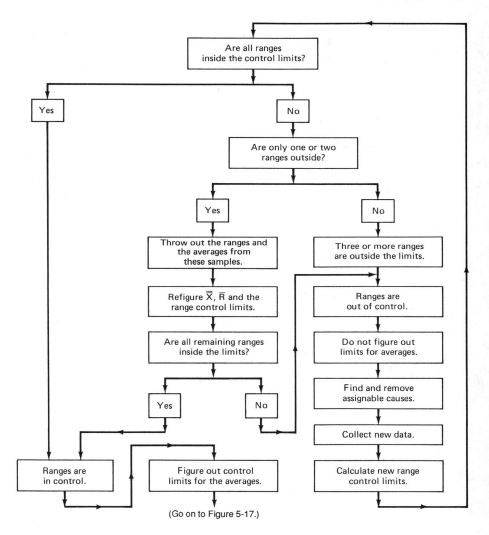

(Go on to Figure 5-17.)

FOR CONTROL LIMITS" to find the A_2 *factor* that corresponds to the sample size you have been using. See Figure 5-15.

Circle the 5 under the "n" column because your samples consist of five double-dip cones. Then circle 0.577 in the A_2 column because this is the A_2 factor to be used with samples of 5. Next, multiply this A_2 factor by 1.35, the average range you found earlier.

$$A_2 \text{ times } \overline{R} \text{ equals } 0.577 \text{ times } 1.35 = 0.779$$

To find the upper control limit, add this figure, 0.779, to the overall average, 4.795, which you found in Step 6.

Figure 5-15. Factors for control limits. A_2 is circled for a sample size of five.

n	A_2	D_4	d_2	$\frac{3}{d_2}$	A_M
2	1.880	3.268	1.128	2.659	0.779
3	1.023	2.574	1.693	1.772	0.749
4	0.729	2.282	2.059	1.457	0.728
⑤	(0.577)	2.114	2.326	1.290	0.713
6	0.483	2.004	2.534	1.184	0.701

Upper control limit for averages equals $\overline{\overline{X}}$ plus (A_2 times $\overline{R}$).
$$= 4.795 + (0.577 \times 1.35)$$
$$= 4.795 + 0.779$$
$$= 5.574$$

To find the lower control limit, subtract 0.779 from the overall average:

Lower control limit for averages equals $\overline{\overline{X}}$ minus (A_2 times $\overline{R}$).
$$= 4.795 - 0.779$$
$$= 4.016$$

In summary, these are the formulas for upper and lower control limits for averages ($\overline{X}$):

Upper control limit for averages $= \text{UCL}_{\overline{X}}$
$$= \overline{\overline{X}} + (A_2 \text{ times } \overline{R})$$

Lower control limit for averages $= \text{LCL}_{\overline{X}}$
$$= \overline{\overline{X}} - (A_2 \text{ times } \overline{R})$$

After checking your arithmetic, draw the control limits on the "AVERAGES" portion of the chart and label them $\text{UCL}_{\overline{X}}$ and $\text{LCL}_{\overline{X}}$. We recommend dashed or colored lines. See Figure 5-16.

Step 13. Are the averages in statistical control?

Make the same kind of check for the averages as you did for the ranges. As with the ranges, there are three possible situations: (1) *all* the averages fall *inside* the control limits; (2) *one or two* averages fall *outside* the limits; or (3) *three or more* averages fall *outside* the limits. (This check for averages applies *only* when you are setting up a new control chart.)

(1) If *all* the averages fall inside the control limits—that is, if no averages fall above the upper control limit ($\text{UCL}_{\overline{X}}$) or below the lower con-

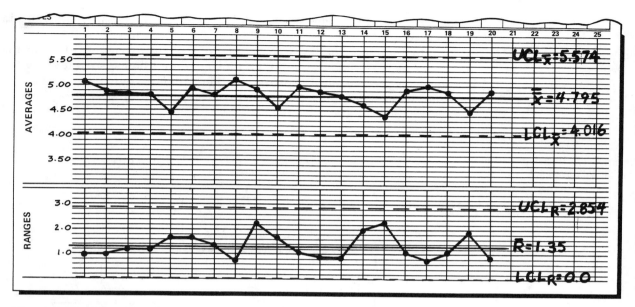

Figure 5-16. Bottom part of average and range chart with control limits for averages.

trol limit (LCL$_{\bar{x}}$)—the averages are in statistical control. Apparently, no assignable causes are disturbing the averages. If the averages and ranges are in control, then you may use your average and range charts to control the ongoing process.

(2) If *one or two* averages fall outside the control limits, it is common practice to throw out those averages for the time being. Then refigure the overall mean ($\bar{\bar{X}}$) and the control limits for the averages without the one or two averages you discarded. If any averages fall outside the new control limits, the averages are out of statistical control. Find and remove the assignable causes. Then, when you think the process is cleaned up, collect new data. You will need to set up *both* the range and the average control charts again.

However, if you refigure the overall mean ($\bar{\bar{X}}$) and the control limits for the averages and find that all the averages are now inside the new control limits, you can use the average and range control charts to control the ongoing process. But be careful—the one or two averages that you threw out could be a signal that some assignable causes are still at work.

(3) If *three or more* averages are outside the initial control limits, the averages are out of control. This situation shows something more than inherent variation. Some assignable cause or causes are at work. Find and remove the assignable causes, start again with new data, and set up new range and average control charts.

120 SPC Simplified for Services: Practical Tools for Continuous Quality Improvement

Figure 5-17 is a decision chart that shows all the possible situations in picture form.

In the chart for the weights of the ice cream (Figure 5-16), all the averages fall inside the upper and lower control limits, so you know the averages are in control. Apparently, there are no assignable causes upsetting the averages. You may use this average and range control chart for the ongoing process.

Figure 5-17. Decision chart for working with averages.

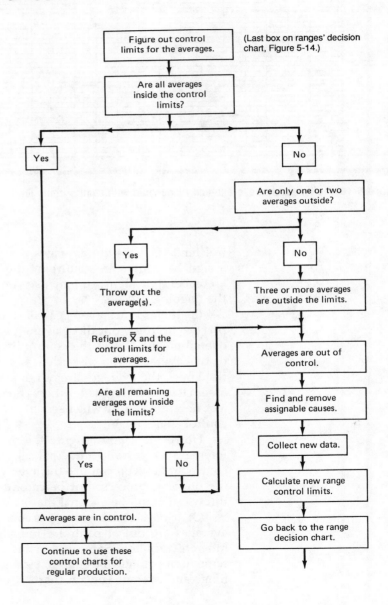

HOW TO USE AVERAGE AND RANGE CONTROL CHARTS IN CONTINUED OPERATIONS

As soon as you find that both the ranges and averages are in control, you can say that the process is in statistical control. This means that, as far as you can tell from the charts, only inherent variation is at work, and that the inherent variation is present because of chance causes. There seem to be no assignable causes. The inherent variation is stable because the range chart is in control. As far as you can tell, the overall process average ($\bar{\bar{X}}$) is also constant because the average chart is also in control.

Once you have set up your average and range ($\bar{X}$-R) control chart limits, you can use them to control regular operations. However, you must remember that the rule about throwing out one or two points outside the control limits applies *only* when you are first setting up the control charts. Once you have established the control limits and start using them in regular operations, a different rule applies: If even *a single point*, either range (R) or average ($\bar{X}$), goes outside a control limit, do not throw out the point. This is a clear indication that an assignable cause is present. You must find the assignable cause, and correct it.

Just because the process is in control, it is not necessarily doing what you want it to do. Being in control simply means that it is humming along smoothly and turning out a *consistent* service. Once the process is in control, you must determine whether or not it is *capable*. (You will learn more about process capability in Module 8). The ranges may be in control, showing that the inherent variation is stable, but the amount of inherent variation may be so large that much of the process is outside the specifications. In this case, you would be faced with a management solvable problem. Management would probably have to redesign the process to decrease the amount of inherent variation, or they might decide to relax the specifications.

You must also determine whether the overall average ($\bar{\bar{X}}$) is adjusted where you want it. The process may be in control, but that is no guarantee that the overall average is where it belongs. The overall average may need to be adjusted so that whatever part of the process you are measuring is within specifications. If you have both upper and lower specifications, you may even decide to adjust the process so that the overall average falls halfway between the specifications.

Now that we have mentioned specifications, we cannot emphasize too strongly the difference between control limits and specifications. *Specifications and control limits have nothing to do with each other*. Specifications are the designer's wishes for individual units—if you are selling shoes, specifications describe the individual shoe; if you sell ice cream cones, specifications describe the individual ice cream cone. In contrast, control limits are based strictly on the variation in the process. They describe the inherent variation in *sample* ranges and averages. In the ice cream problem, you will notice that we never gave you the specifications.

If the inherent variation is small enough so that the process can meet specifications, continue the process while using your control charts. If not, then you may have to take the problem to management. How do you know whether the inherent variation is small enough? You will probably have to run a process capability test, which we will describe in Module 8. You may also have to adjust the process so that the overall average ($\bar{\bar{X}}$) is where you want it. This adjustment should not affect the control limits for the ranges, but it will change the limits for the averages.

We recommend that once you have made this adjustment, you check it out by looking at the next few averages. Take 10 new samples as close together as possible. Calculate the overall average of these samples and draw it in on your control chart for averages. Figure out new upper and lower control limits by using the new overall average, but keep the old average range:

$$\text{new UCL}_{\bar{X}} \text{ equals new } \bar{\bar{X}} \text{ plus } (A_2 \text{ times old } \bar{R})$$

$$\text{new LCL}_{\bar{X}} \text{ equals new } \bar{\bar{X}} \text{ minus } (A_2 \text{ times old } \bar{R})$$

Now you should be ready to use the control charts, as we discussed in the first section of this module.

MEDIAN AND RANGE CHARTS

Now that you have learned how to make and use average and range charts, let's take a look at another type of control chart, which you can sometimes use in place of an average and range chart. This is the *median and range* ($\tilde{X}$-R) *chart*.

The median and range chart is easier to use than an average and range chart, but is not suitable in all situations. It is a good chart to use when you know that the process for delivering or producing a service (1) follows a normal (bell-shaped) distribution, (2) is not very often disturbed by assignable causes, and (3) can be easily adjusted by the employee. If the process does not meet these requirements, you should use an average and range chart.

Developing the median and range chart is similar to developing the average and range chart. It is easy to use, once you have established the control limits. For a median and range chart, you can use a sample size of 2 to 10, but a sample of 3 or 5 is the easiest to work with.

The form for this chart looks very much like the average and range chart form. In fact, you can use the average and range chart form. Just label it as a median and range chart. Cross out the words that don't apply and write in the ones you need. See Figure 5-18.

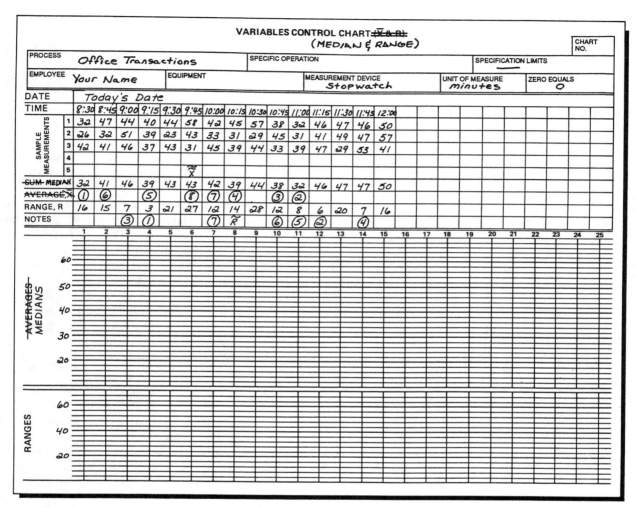

Figure 5-18. Median and range ($\tilde{X}$-R) chart with samples, medians, ranges, and scales.

DEVELOPING A MEDIAN AND RANGE CHART

The chart shown in Figure 5-18 is based on a three-piece sample. The procedure for developing the chart is as follows.

Step 1. Collect samples.

Take samples that were made as close to the same time as practical.

Step 2. Record the measurements.

Take your measurements and record the results in the "SAMPLE MEASUREMENTS" columns.

Step 3. Determine the median measurement.

Find the *median* ($\tilde{X}$) measurement of the sample and record it in the "MEDIAN" row on the chart. A median is always in the *middle* of a group of measurements when you count from the smallest to the largest measurement. Half the measurements will be smaller than the median and half will be larger.

In this chart, you are using a three-piece sample, so the median is the measurement between the largest and smallest measurements. If you had chosen to use a five-piece sample, the median would be the third-largest measurement, which falls between the two largest and the two smallest measurements. If you must use an even-numbered sample size, the median is the number halfway between the two middle measurements in the sample (for example, between the third-largest and fourth-largest pieces of a six-piece sample). An easy way to figure the median in a sample with an even number of pieces is to add the two middle measurements in the sample and divide by 2. This result is the median and will be halfway between these two measurements. We recommend that you stay with the odd-numbered sample size, because it is easier to use.

Do not confuse the *median* of a group of measurements with the *mean* of that group. The mean is the average value of the measurements and the median is the middle measurement of the group. In a group of five measurements, 7, 3, 8, 6, 10, for example, the average is 6.8 (7+3+8+6+10=34 divided by 5, which equals 6.8). To find the median, first rank the numbers by size: 3, 6, 7, 8, 10. There are two numbers below the 7 and two numbers above the 7, so 7 is the middle number, or the median.

Step 4. Determine the ranges for the samples.

Figure the range, as you have already learned to do, by subtracting the smallest measurement from the largest. Record the range readings in the "RANGE" row on the chart.

In Figure 5-18, the first three-piece sample contains the measurements 32, 26, and 42. When you rank them (26, 32, 42), you can see that 42 is the largest and 26 is the smallest, so the median is 32, the number between these two measurements.

To figure the range of the first sample, subtract the smallest measurement (26) from the largest (42). The result, 16, is the range.

As you can see on the chart, the second sample is

> 47 (largest)
> 32 (smallest)
> 41 (median)
> 47 (largest) minus 32 (smallest) equals 15 (range).

Step 5. Determine the median of medians ($\tilde{\tilde{X}}$) and the median of ranges ($\tilde{R}$).

When you have measured and recorded 15 samples, you figure the *median of medians* ($\tilde{\tilde{X}}$) and the *median of ranges* ($\tilde{R}$).

To find the median of medians, count up from the smallest median to find the middle value of the 15 medians. The middle value in the group of 15 is the eighth median from the bottom (or the top) of the values. On the chart in Figure 5-18, we have ranked the eight smallest medians by using circled numbers in the row below the medians to help you count up from the smallest to the eighth in the group of 15. The first and smallest median is 32; the next smallest is also 32, and is numbered 2; and the next is 38 and is numbered 3; the next is 39 and numbered 4; number 5 is another 39; number 6 is 41; number 7 is 42; and number 8 is 43. This number, 43, is the median of the medians ($\tilde{\tilde{X}}$). For the time being, write $\tilde{\tilde{X}}$ over this value.

Find the median of ranges or the median range in the same way as you found $\tilde{\tilde{X}}$. Count the 15 range values from the smallest to find the eighth, or middle, value. The chart in Figure 5-18 shows that the median range value ($\tilde{R}$) is 14. Write $\tilde{R}$ underneath the 14.

Step 6. Set the scale for the median ($\tilde{X}$) chart.

Select a scale for the median ($\tilde{X}$) chart such that the spread between the largest and smallest of all the individual measurements will take up about one-half to three-fourths the available chart space on the median chart.

Because the median of the medians ($\tilde{\tilde{X}}$) is 43 in this example, set the value 40 at the middle of the scale. Then each major division on a heavy line will be worth 10. The scale we have marked runs from 20 to 60. Write these numbers on the chart in the area marked "MEDIANS."

Step 7. Set the scale for the range.

Do this the same way as you set the median chart scale.

Each major division on the range chart scale is worth 20 units. (The divisions have been marked along the side of the range chart.) This arrangement allows room for the upper control limit on the chart.

Step 8. Plot the measurements on the control chart.

All the measurements in the sample are plotted on the median and range chart. For the first sample, place small dots on the vertical line at the appropriate place on the scale. Place the first measurement at 32, the second at 26, and the third at 42. Identify the dot representing the median, which is 32, by placing a small circle around that dot. Plot the rest of the measurements in the same way. See Figure 5-19. Connect the medians with straight lines to help you spot trends. Draw in a line for $\widetilde{\widetilde{X}}$, the median of medians, and label it.

Plot the ranges on the median and range chart just as you did on the average and range chart. Draw in a line for $\widetilde{R}$, the median of ranges, and label it.

Step 9. Determine the control limits for medians and ranges.

Calculate the control limits for the medians ($\widetilde{X}$) and the ranges (R) by using formulas and factors that are similar to the ones for average and range charts. The factors are shown in Figure 5-20. Note that the $\widetilde{A}_2$ and $\widetilde{D}_4$ factors for the median and range chart are *not* the same as the A_2 and D_4 factors for the average and range charts.

To figure the upper control limit (UCL_R) for the ranges, multiply the median range (14) by the $\widetilde{D}_4$ *factor* from the chart in Figure 5-20. This value is 2.75 for a sample size of 3. The calculation is as follows:

$$UCL_R \text{ equals } \widetilde{D}_4 \text{ times } \widetilde{R}$$
$$UCL_R = 2.75 \times 14 = 38.5$$

The lower control limit for ranges (LCL_R) is zero.

At this point, check all the range numbers on the chart to see whether any are larger than 38.5. If *no more than two* ranges fall outside the upper control limit, throw out those ranges and their medians. Find the new median range and refigure the upper control limit. Now should *any* ranges fall outside this new control limit, you must start over with new measurements. If *three or more* ranges fall outside the original control limit, start over with new measurements.

In the example, no ranges fall outside the upper control limit, so now you can figure the control limits for the medians.

The $\widetilde{A}_2$ *factor* for a sample size of 3 is 1.26. See Figure 5-20. Multiply this number by the median range ($\widetilde{R}$) and add the result to the median of the sample medians ($\widetilde{\widetilde{X}}$) to find the upper control limit for medians ($UCL_{\widetilde{X}}$).

$$UCL_{\widetilde{X}} \text{ equals } \widetilde{\widetilde{X}} \text{ plus } (\widetilde{A}_2 \text{ times } \widetilde{R}).$$

First, multiply $\widetilde{A}_2$ by $\widetilde{R}$. Then add that to $\widetilde{\widetilde{X}}$, which is 43.

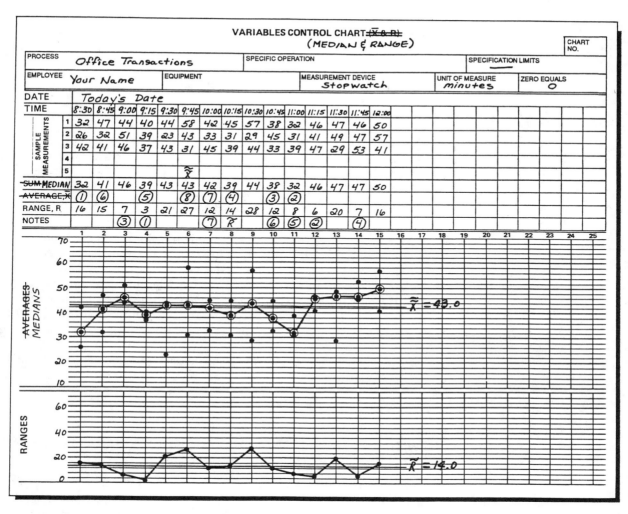

Figure 5-19. Median and range chart with points plotted.

Figure 5-20. Control limit factors for median and range charts.

Sample size, n	$\tilde{A}_2$	$\tilde{D}_4$
2	2.22	3.87
3	1.26	2.75
4	0.83	2.38
5	0.71	2.18

$$UCL_{\tilde{X}} = 43 + (1.26 \times 14)$$
$$= 43 + 17.64$$
$$= 60.64$$

The lower control limit for medians ($LCL_{\tilde{X}}$) is as follows:

$LCL_{\tilde{X}}$ equals $\tilde{\tilde{X}}$ minus ($\tilde{A}_2$ times $\tilde{R}$)

First, multiply $\tilde{A}_2$ by $\tilde{R}$. Then subtract the result from $\tilde{\tilde{X}}$, which is 43.

$$LCL_{\tilde{X}} = 43 - (1.26 \times 14)$$
$$= 43 - 17.64$$
$$= 25.36$$

Again, compare the medians to the control limits to see whether any medians fall outside the upper or lower control limits. If *one or two* fall outside, throw them out, refigure the median of the medians, and then calculate new control limits. If *any* medians fall outside the new control limits, throw out the chart and start over with new measurements. If *three or more* medians fall outside the original control limits, throw out all the measurements and start over, just as you did with the average and range charts.

Draw control limits using broken or dashed lines, as shown on the chart. See Figure 5-21. It's good practice to identify the lines with their values, as shown on the chart.

HOW TO USE MEDIAN AND RANGE CONTROL CHARTS IN CONTINUED OPERATIONS

Once you have set up your median and range ($\tilde{X}$-R) control chart limits, you can use them to control regular operations. Remember, you may throw out the one or two points falling outside the control limits *only* when first setting up the control charts. Once you have established the control limits and start using them on a regular basis, a different rule applies: If even *one* point, either range (R) or median ($\tilde{X}$), goes outside a control limit, you know an assignable cause is present. Find and correct it.

The plotted points on the median chart are individual measurements. For this reason, you can compare these points to the specifications. But remember, you must take corrective action when the median values are outside the control limits. Individual points (not medians) can appear outside the control limits even when the process is behaving normally. Individual points outside the specifications do not indicate that the process is out of control. This is not a symptom of an assignable cause. Remem-

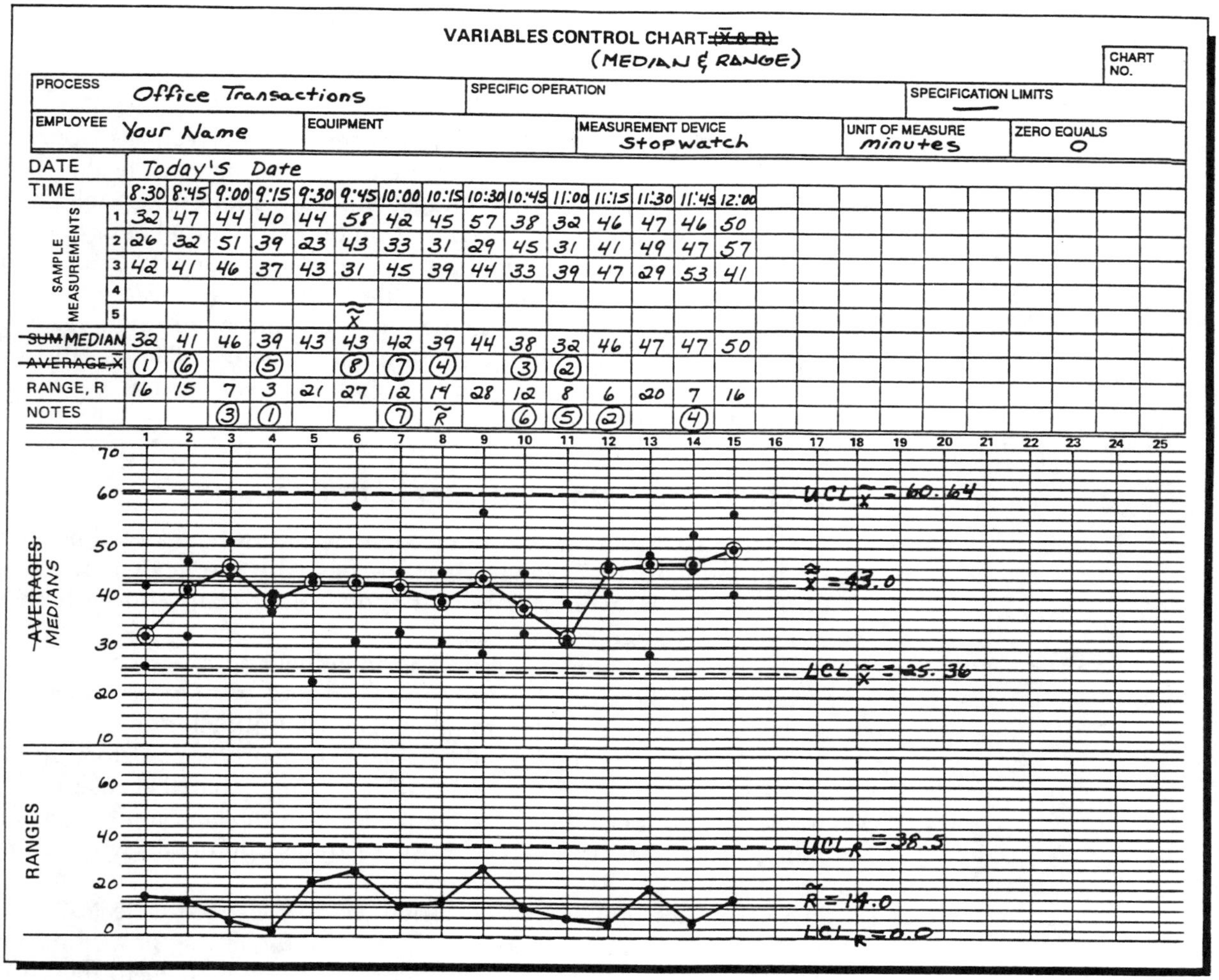

Figure 5-21. Median and range chart with control limits.

ber, the process needs to be adjusted *only* when medians ($\tilde{X}$) or ranges (R) lie outside the control limits.

INDIVIDUAL AND RANGE CHARTS

An *individual and range (X-R)* chart can be useful in special situations. This control chart is based on individual measurements rather than small samples.

The individual and range chart does not detect changes in the process as quickly as an average and range chart. Like the median and range chart, it is best to use the individual and range chart when we already know that the frequency distribution of the measurements from the process of producing or delivering a service matches the bell-shaped (normal) curve.

A good place to use this chart is whenever you can take only *one* measurement or reading. For example, a private medical facility could track how much under or over budget it is each month. A public utility could use this type of chart to analyze or monitor the number of customer complaints per month. A school district could monitor total energy consumption in its buildings. In each of these situations, you are only able to take one reading during a time period. Since these are single readings, an individual and range chart is appropriate. You can also use the individual and range chart when you find that it's costly or time consuming to test or take readings.

DEVELOPING AN INDIVIDUAL AND RANGE CHART

The standard average and range chart form can be adapted easily for use as an individual and range chart. Just as with the median and range chart, cross out the words that don't apply to this chart and write in the words you need. See Figure 5-22.

To demonstrate how to set up an individual and range chart, we will use miles per gallon from a maintenance record of one car in a fleet of leased vehicles. If you wanted to monitor or control fuel usage, you would find that a fairly long time would elapse between fill-ups before you obtained 4 or 5 readings for an average and range chart because there is only one reading per fill-up. To overcome this problem, you can develop an individual and range chart with the readings you obtain when you fill the gas tank.

To set up an individual and range chart, you need at least 20 readings. We are using 25 for this example. Record the miles-per-gallon values on the control chart, as shown in Figure 5-22. Enter them in the "SAMPLE MEASUREMENTS" row. Remember this is an individual chart, so you will have only one reading. On this chart, we record the date of the fill-up in the "DATE" row to show the order of the samples.

To figure the range (R) values, calculate the differences between the individual measurements. The first range is the difference between the first and second measurements (28.5 minus 25.5 equals 3.0). Record it in the "RANGE" row on the chart under the second miles-per-gallon entry, 25.5. There is no range value for the first entry, so there will always be one less range value than there are individual readings. See Figure 5-22.

Set the scale for the "individuals" chart so that the difference between the largest and the smallest readings will take up about half the space on the chart.

The "ranges" chart will start at zero. Each division on the scale should

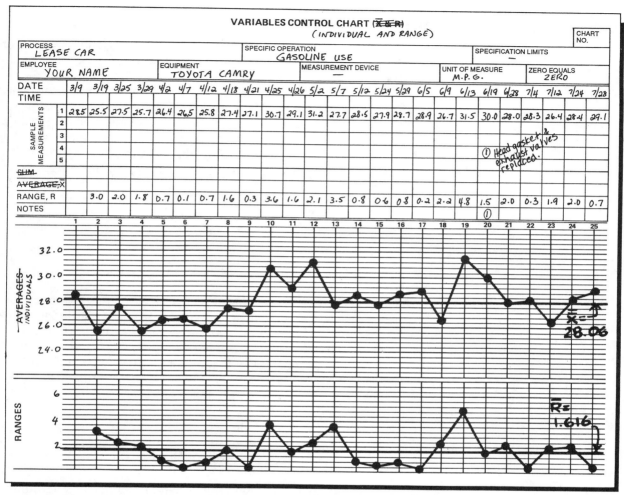

Figure 5-22. Individual and range (X-R) chart with miles-per-gallon values, ranges, and points plotted on graph. Lines are drawn for $\bar{\bar{X}}$ and $\bar{R}$.

have the same value as a division on the "individuals" chart. (In this chart, each division equals 0.4 miles per gallon, so five divisions represent 2 miles per gallon.)

CONTROL LIMITS FOR INDIVIDUAL AND RANGE CHARTS

Figure control limits in the same way you figured control limits for average and range charts. The upper control limit for ranges is calculated in this way:

$$UCL_R \text{ equals } D_4 \text{ times } \overline{R}.$$

The D_4 factor is the same as you use on an average and range chart when the sample consists of two pieces. This value is 3.268. (See Figure 5-12; n=2.)

The value of $\overline{R}$ (the average range), as you already know, is calculated by adding all the ranges and dividing by the number of range values. In this case, the total of the ranges is 38.8 and the number of range values is 24. (Remember that on the individual and range chart, there is always one less range than the number of individual readings.)

On this chart (Figure 5-23), the upper control limit for ranges is calculated as follows:

$$\overline{R} = 38.8 \div 24$$
$$\overline{R} = 1.616$$

$$UCL_R = D_4 \overline{R}$$
D_4 times $\overline{R}$ equals 3.268 times 1.616 equals 5.28.

The lower control limit for ranges is always zero on this type of chart.

To calculate the average of the individuals ($\overline{\overline{X}}$), add all the individual readings and divide by the number of individual readings. On this chart (Figure 5-23), the sum of the individual readings is 701.5 and there are 25 readings. The control limits for individuals are calculated as follows:

$$\overline{\overline{X}} = 701.5 \div 25 = 28.06$$

Upper control limit for individuals, UCL_X:

$$UCL_X \text{ equals } \overline{\overline{X}} \text{ plus } (2.66 \text{ times } \overline{R}).$$

Be sure to multiply 2.66 times $\overline{R}$ *first* before adding $\overline{\overline{X}}$. (The number 2.66, which is used to calculate the control limits, has been determined by mathematicians.)

$$UCL_X = 28.06 + (2.66 \times 1.616) = 32.35$$

Lower control limits for individuals, LCL_X:

$$LCL_X \text{ equals } \overline{\overline{X}} \text{ minus } (2.66 \text{ times } \overline{R}).$$

Be sure to multiply 2.66 times $\overline{R}$ *first* before subtracting the result from $\overline{\overline{X}}$.

$$LCL_X = 28.06 - (2.66 \times 1.616) = 23.76$$

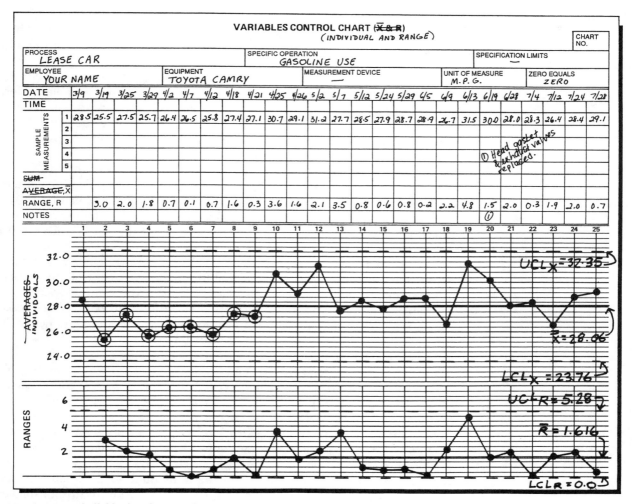

Figure 5-23. Individual and range chart with control limits.

When you have calculated the upper control limit for the range, look at all the range values to see whether any of them are greater than the upper control limit. Remember: (1) If *none* of the ranges is above the upper control limit, go ahead and calculate the control limits for the individual readings. (2) If *one or two* ranges are greater than the upper control limit, eliminate them from the group of range values and calculate a new average range ($\bar{R}$) and a new upper control limit for ranges. Be sure to throw out the corresponding individual readings (X's). Now if *any* of the remaining range values are still greater than the new upper control limit, throw out all the readings and start over with new data. But if *none* of the remaining range values is greater than the new upper control limit, go ahead and calculate the upper and lower control limits for the individual readings. (3) If *three or more* range values are greater than the control limit, throw out all the readings and start over.

Apply the same tests to the individual readings. If *three or more* readings lie outside the control limits, throw out the readings and start over. If *one or two* lie outside, refigure the control limits for individuals and check to see whether any more readings are still outside the control limits. If so, start over with new readings. When no readings lie outside the control limits, go ahead and use the chart.

As you can see in Figure 5-23, the range values and the individual readings in the example all fall within the control limits. However, there are eight individual points in a row below the $\overline{\overline{X}}$. Remember that this is another indication of the presence of assignable causes. Maintenance should look for these causes, remove them, and construct a new chart with new data.

Once you have established the individual and range chart with proper control limits, you can use it as a tool to monitor and control the process of producing or delivering the service. Use it in the same way as other variables control charts. But since this is a control chart of *individual* readings, you may compare the readings you plot on the individual chart to the specification limits as well as to control limits.

When even one point on the control chart falls outside the control limits, you have a problem that employees can solve. Look for assignable causes and take corrective action. When the points on the control chart fall outside the specification limit but *not* outside the control limits, you have the type of problem management must solve. In this case, you are doing things as planned and instructed, but the readings are out of specification. To correct this situation requires a change in plans, instructions, or process.

If either of the control limits for the individual readings falls outside the specification, you can predict that the process will produce or deliver the service outside the specification some of the time. For this reason, be sure your prediction is accurate. To gain confidence in the accuracy of the control limits, it is good practice to recalculate the control limits as soon as you obtain more readings.

When you have added 20 more readings to the control chart, add them to the original group and refigure the control limits. Do this until you have calculated the control limits using at least 100 readings.

SUMMARY

Control charts can tell you either to leave your process alone and make no adjustments, or to adjust or correct the process. In the first case, only inherent variation appears to be at work. In the second case, an assignable cause has probably disturbed the process.

Use an average and range control chart when you have variable data (data based on readings).

To set up average and range charts, you must collect samples; calculate $\overline{X}$'s and R's; calculate $\overline{\overline{X}}$ and $\overline{R}$; determine the control limits for ranges and know how to use them; and determine the control limits for averages and know how to use them. You must check to see whether the ranges are in statistical control before you calculate the limits for averages. A decision chart will help you to work with ranges and averages.

Points inside the control limits mean that only inherent variation is at work and no adjustments are required in the process. Points outside the control limits, as well as certain patterns within the limits, indicate assignable causes. In that case, you must track down the cause—equipment, material, method, environment, or people—and correct it.

When the process is in statistical control, determine whether the inherent variation is small enough so that the process of producing or delivering your service can meet specifications. Remember that control limits have *nothing* to do with specifications.

A range chart tells you whether or not the process variation is stable. If it is stable, then you can check to see how well the service you are producing or delivering is meeting specifications. You will learn more about this in Module 8.

The median and range chart is easier to use than the average and range chart. The median is the middle value in a group of numbers, and can be found with little or no arithmetic. The range is easy to calculate because the points on the chart show clearly which is the largest and which is the smallest measurement. Sample sizes of three or five are easiest to work with.

You can use the median and range chart when you know that the frequency distribution of measurements from a process matches the normal, or bell-shaped, curve. This is not as important when using the average and range chart.

The individual and range chart can be used in special circumstances, such as when you can take only one measurement or reading or when a long time elapses between readings.

On the individual and range chart, you record individual readings, and you can compare the points in the individual chart to the specifications. This chart is easy to use, but it does not detect changes in the process as quickly as the average and range chart. The most important condition for the use of this chart is that the frequency distribution of the readings from the process you want to chart should match the normal bell-shaped distribution.

PRACTICE PROBLEMS: VARIABLES CHARTS

Work through the following problems using the statistical techniques you learned in Module 5. Solutions can be found in the "Solutions" section beginning on page 259.

Problem 5-1.

Look again at the data for the weights of ice cream in Problem 4-1. Construct average and range charts using the following groupings of data. The sample size is five.

1. A,B,C,D.
2. E,F,G,H.
3. I,J,K,L.
4. M,N,O,P.
5. A,E,I,M.

(a) Do you see any differences among the overall averages ($\bar{\bar{X}}$'s)?
(b) How do the overall averages compare to the estimates of the process averages you made using the frequency histograms?
(c) How does the fifth control chart (A, E, I, M) compare to the other four?

Problem 5-2.

A radio station is testing some of its radio equipment. The unit of measurement is called the decibel (db). Technicians take five readings every 4 hours, so the readings shown are in the order they were taken. Construct a median and range chart. Divide the readings into subgroups of five. You will have a total of 20 median points on the chart.

(a) Is this process in control?
(b) Are any individual values outside the control limits? What action, if any, should you take?

db readings

28, 28, 21, 29, 29
35, 28, 31, 31, 28
24, 33, 29, 30, 28
25, 28, 27, 28, 30
23, 30, 30, 32, 31
27, 30, 28, 28, 29

db readings (cont.)

30, 26, 33, 27, 25
30, 26, 30, 33, 34
24, 29, 26, 25, 26
29, 28, 30, 27, 26
28, 28, 28, 32, 24
32, 30, 25, 29, 31
35, 28, 28, 26, 29
29, 23, 29, 31, 30
31, 26, 33, 28, 28
34, 27, 31, 26, 31
28, 30, 29, 27, 27
30, 33, 39, 29, 25
31, 29, 31, 27, 36
34, 29, 34, 28, 30

Problem 5-3.

The book acquisitions department of a university library is working with a local company to upgrade the library's holdings in humanities, engineering, and science. The library feels it should be ordering between 200 and 400 books each month. Someone in the department has tracked book orders for 23 months. Construct an individual and range chart from these data.

(a) Why should you use an individual and range chart for this problem?
(b) Is the process for ordering books meeting specifications?
(c) Is the process for ordering in control?

Number of Books Ordered by Month		Number of Books Ordered by Month	
Jan.	275	Jan.	368
Feb.	335	Feb.	325
Mar.	336	Mar.	400
Apr.	363	Apr.	491
May	319	May	500
Jun.	400	Jun.	400
Jul.	376	Jul.	175
Aug.	245	Aug.	297
Sep.	240	Sep.	170
Oct.	300	Oct.	271
Nov.	210	Nov.	250
Dec.	363	Dec.	—

MODULE 6
Attributes Control Charts

NEW TERMS IN MODULE 6
(in order of appearance)

attributes

attribute or counting data

p-chart

np-chart

c-chart

percent defective p-chart

$\bar{p}$ *(average percent defective for the process)*

$\sqrt{}$ *(square root)*

fraction defective, or proportion defective, p-chart

$n\bar{p}$ *(average number defective for the process)*

inspection unit

$\bar{c}$ *(average number of defects for the process)*

When you are doing quality control work, you will be using or making charts based on two kinds of information. In Module 5, you learned to use and construct average and range and other kinds of control charts. The type of information you need for those charts is called *variable data*. Remember that when you weighed the double dips of ice cream, you found that the weights varied from one cone to the next. Variable data come from things you can weigh or measure, such as ounces of ice cream, or degrees Centigrade.

Sometimes, though, you have a different situation when you are producing or delivering a service. The process or service has a characteristic you want to control, but you cannot measure it or it is difficult to measure. Yet the characteristic is important because of cost or acceptability to the customer.

When you're controlling for these characteristics, inspection or sampling shows whether the service meets standards, conforms to specifications, or has one or more defects or errors. It's either O.K. or not O.K. One example is when a supervisor checks whether or not a room is ready for the next guest. Such nonmeasurable characteristics are called *attributes*. They provide *attribute or counting data*.

WHY USE AN ATTRIBUTES CONTROL CHART?

Whenever you need to monitor a nonmeasurable characteristic in your process or service, you can use a *p-chart*, an *np-chart*, or a *c-chart*. The p-chart helps you monitor and control the percentage of defective units in the process or service. The np-chart helps you monitor the number of defective units. The c-chart helps you monitor the number of defects in the items. If you want to monitor the percentage of defective units in an operation, use a p-chart. If you want to monitor the number of defective units, use an np-chart. If you want to monitor the number, or count, of defects in an element of a service, use a c-chart. You can remember which is which if you keep in mind that "p" stands for "*p*ercentage," "np" stands for "*n*umber of defective units," and "c" stands for "*c*ount."

We are using the word "defective" to describe a part or unit of a service. The unit is either good (acceptable), or bad (not acceptable), that is, defective. Also, we are using the word "defect" in a special way. There may be one or more defects or errors in the unit of the service, but the entire unit is defective whether it has one defect or more than one.

In this module, you will learn to use and construct all three kinds of attributes charts, beginning with the p-chart.

PERCENT DEFECTIVE p-CHARTS

Most p-charts use percentages. A percentage is the number of units out of 100 units that are defective or otherwise do not conform to specifications. We call this kind of p-chart a *percent defective p-chart*.

To find the percentage, p, use this formula:

> p equals the number of defective units divided by the sample size. Multiply the result by 100%

In a sample of 100 units where seven units are defective,

$$p = 7 \div 100 \times 100\% = 7\%$$

On a percent defective p-chart, you plot points as percentages of defective units in the samples.

HOW TO USE p-CHARTS

As you did in Module 5, look first at a percent defective p-chart that somebody else has set up. The supervisor developed the chart based on several weeks' work. She continued sampling the credit summaries on July 21 and plotted the results as shown in Figure 6-1.

Your job is to take the next sample, calculate p, enter the data on the

Figure 6-1. p-chart already set up. (Form developed by Robert and Davida Amsden and Howard Butler.)

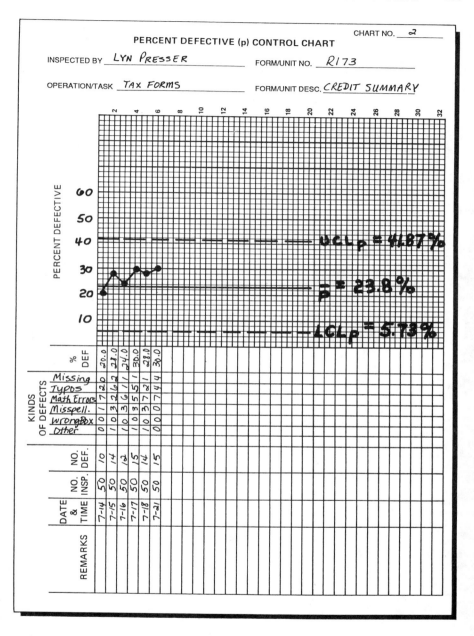

p-chart, and plot the point. You will also have to understand what the chart is telling you, so that you can determine whether to continue the review process or whether to take corrective action.

In this case, let's suppose you're checking a tax form called Form R173. Your sample is 50 forms. You must check the way each form has been filled out to determine whether it's O.K. or not O.K., regardless of the number or type of defects or errors you find.

Since you will be inspecting the forms and filling in the chart for the day, you would write your name next to Lyn's. See Figure 6-2. Write in 7/22 under the "DATE & TIME" heading for the seventh sample (your first) and write your name in the "REMARKS" row for 7/22. Check the

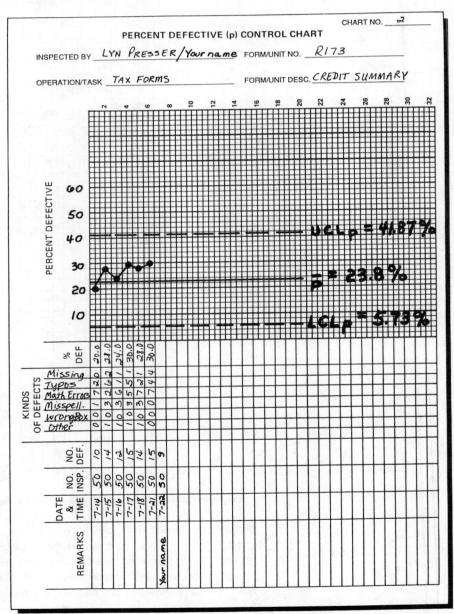

Figure 6-2. p-chart with new data entered for 7/22.

next 50 forms. You find nine defective forms. Under the heading "NO. INSP." (number inspected) write 50; under "NO. DEF." (number defective) write 9.

You probably have noticed divisions under the heading "KINDS OF DEFECTS." You may also be asked to fill in the number of the particular kinds of defects such as "missing," "typos," "math errors," "misspell.," "wrong box," and "other." To fill these in, count how many of each type of defect you find and write that number under the appropriate heading. Remember that a form may have more than one kind of defect. It might have typos *and* math errors, for example. But even though a form has *more than one error*, it counts as only *one defective form*, called a "defective."

In this sample of 50, one form has a misspelling, eight forms have math errors, and one has a typo. (One of the defective forms had two types of defects. It had a typo and a math error. The number of defective forms is 9 although the number of errors is 10.) Fill in the appropriate boxes under "KINDS OF DEFECTS." (See Figure 6-3.)

Once you have filled in the data, you are ready to calculate the percent defective in this sample of 50. Remember the formula:

p = number of defectives divided by sample size times 100%.
p = 9 ÷ 50 × 100% = 18.0%

Write 18.0 under "% DEF" (percent defective) in the 7/22 column. See Figure 6-3.

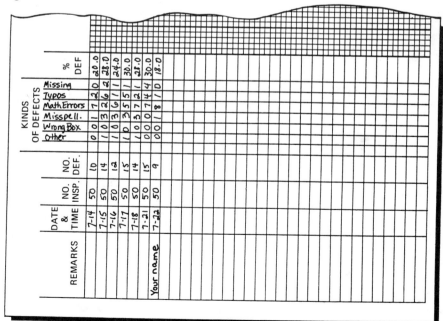

Figure 6-3. Portion of p-chart showing kinds of defects and percent defective for 7/22.

Now plot the point as a large dot on the graph and draw a line connecting this dot with the last one. See Figure 6-4.

As you remember from Modules 1 and 5, averages and ranges have upper and lower control limits, UCL and LCL, respectively. The same is

Figure 6-4. p-chart showing p plotted for 7/22.

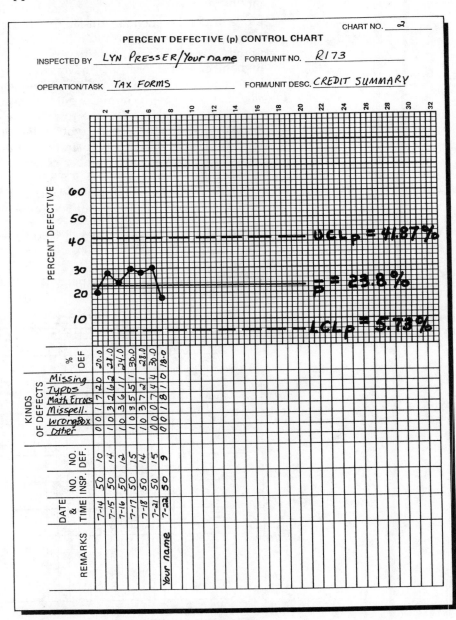

Module 6: Attributes Control Charts   143

true for p-charts. Keep this in mind when you look at what you plotted. You will learn how to set control limits later in this module.

Is the p where it's supposed to be? It's 18%, and that falls between the upper control limit of 41.87% and the lower control limit of 5.73%, so it's all right.

What is this point telling you? When the percent defective is within the control limits, your process is in statistical control. The rule is: If the p, percent defective, for the sample is within the control limits, keep running the process without correction.

INTERPRETING PERCENT DEFECTIVE p-CHARTS

Once you know how to figure the percent defective for the samples and how to plot the points, you can go on to interpret the charts.

PERCENT DEFECTIVE, p, INSIDE CONTROL LIMITS

Now that you have been taking the samples for a while, look at your p-chart in Figure 6-5. Since every point is inside the control limits, what should you do? Keep running the process. When all points are inside the limits, no corrections are needed. (If any point lies directly *on* the upper or lower control limit, consider that point as inside.)

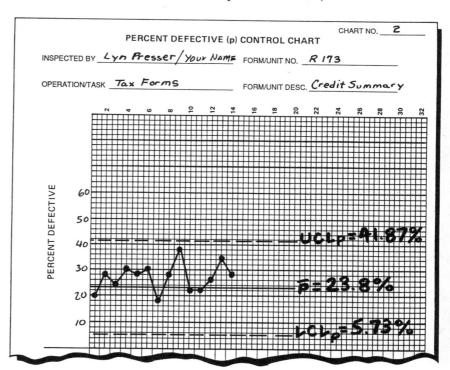

Figure 6-5. Portion of p-chart showing p inside control limits.

What do you do when a point is very close to the upper or lower control limit? If the point is inside or on the limit, make no corrections and keep running the process. But if the point is outside, an assignable cause is at work. Find that cause and correct it.

We have said before that a process is in statistical control when all points are inside the control limits, even if the points move up and down as they do in Figure 6-5. This movement most likely shows the natural variation in the process. As you learned in Module 5, this kind of variation is called inherent variation and is due to chance causes. Such causes are usually beyond the employee's immediate control, but they do cause the points to move up and down inside the control limits.

PERCENT DEFECTIVE, p, OUTSIDE CONTROL LIMITS

Figure 6-6 shows a p outside the control limits. Because it is outside, this process is now out of statistical control. An assignable cause is present, and something more than simple inherent variation is at work.

As you recall from Module 5, an assignable cause isn't always present. It's not natural to the process. But when it shows up on the control chart, you may be able to track it down.

Often a point outside the limits indicates a problem you can handle on the job, such as forgetting to check for misspelled words by using a dictionary. At other times, however, such as when there is a new reporting procedure, your boss must step in. Still, the important thing is not *who* is responsible, but *what* has changed and what needs to be done. The point outside the control limits simply shows you that a *change* has taken place and *when* it took place.

What do you do? The chart is telling you to correct your process. For some reason the process has shifted upwards, and it needs to be adjusted down.

In the case of Tax Form R173, the percent defective may have gone outside the limit because some information was missing or had not been collected properly; or there were typos because keys on the keyboard needed to be cleaned and were sticking.

Figure 6-6. Portion of p-chart showing p outside control limits.

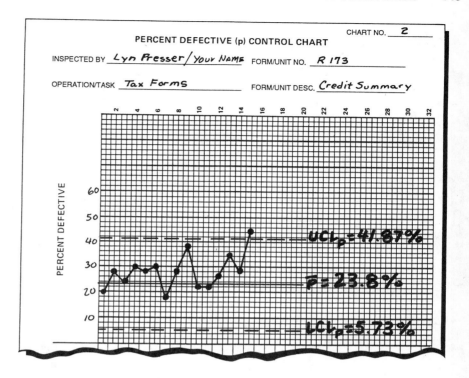

OTHER INDICATIONS OF OUT-OF-CONTROL PROCESSES

As you recall for Module 5, the points you plot may show a pattern. One pattern we mentioned is seven consecutive points above or below the line marked $\bar{p}$. (We'll explain the $\bar{p}$ later.) Even though all the points are still inside the control limits, this pattern is giving you an indication— a weaker one—that the process may be out of control.

Other patterns include longer runs of points, as well as too many or too few points close to the $\bar{p}$ line. To interpret these patterns, we suggest that you look them up in the books we recommended in Module 5 on page 105.

TYPES OF ASSIGNABLE CAUSES

The point or points outside the control limits are your clearest indication that an assignable cause is at work in your process. Let's review the sources of assignable causes. (See Module 1, Figure 1-4, for the fishbone diagram.)

(1) Check the *equipment*. Has something about the equipment itself

changed so that points are outside the limits? Look for wear, settings, the need for maintenance, and so on.

(2) *Materials* are another item to check. When an assignable cause is due to materials, something is different about what is coming to the process. There may be a change in the reservation forms, for example, or the meat supplier has changed the fat content in the ground beef.

(3) Another area to check is *method*. By "method" we mean the procedure followed in filling out the form, such as the way you handle state taxes.

(4) Also look at the *environment*, which includes such things as lighting, height of the work table, and humidity or dryness.

(5) Finally, consider the *employee*. Check training, hearing, eyesight, arithmetic errors, and fatigue.

SETTING UP PERCENT DEFECTIVE p-CHARTS

Now you can try setting up your own percent defective p-charts for one of several purposes. (1) You may be using the charts for control. That is, they can help you determine whether to continue the process "as is" or whether you have to make corrections. (2) You could be using the charts for analysis to look for differences between materials, days, or shifts. (3) The charts may help in communication and documentation. Remember, charts can help focus attention on consistent quality.

We will use the example of the credit summary form (R173) to show you how to set up p-charts. Corporate quality control and office management agree that it's necessary to control the percent defective in the credit summaries. You're the employee. The p-chart will help you control your process, and will also show you, your supervisor, and the corporate people how well your process is doing.

Step 1. Take the sample.

Here are some guidelines for determining sizes of samples for p-charts:

- Use a sample size of at least 50 units.
- Use a sample size big enough to give an average of four or more defectives per sample.
- Avoid very large samples taken over a long period of time. If possible, break the sampling time period into smaller segments. Instead of plotting the percent defective for a day's production, break the day into 2-hour or 4-hour time periods and plot the percent defective for these shorter periods.
- If the sample sizes vary and the sample is more than 20% greater or 20% less than the average sample size, calculate separate control limits for that sample.

You want your sample to show only inherent variation. Remember, inherent variation means that defects in the forms are showing up at random. They're due to chance causes. If your sample shows only ordinary (inherent) variation, you can draw accurate control limits.

In order to have only inherent variation, you need to take your samples over a short period of time from a single source. This means you will sample one type of form, one word processor, and one employee.

Now you're ready to begin taking your sample. The advisor from corporate quality assurance has established that the best sample size for p-charts for checking the revised Form R173A is 50 units. The process has been running at about 28% defective, so a sample of 50 should include about 14 defectives (28% of 50 is 14). Therefore, a sample of this size should easily meet the guideline that calls for an average of at least four defectives per sample.

Take 50 Credit Summary Form R173As from your work station, one after the other as they are completed. You are the only employee, and you're taking a sample of 50 once during the day. As far as you can determine at this point, every sample of 50 forms is like every other sample. Any variation from sample to sample should be no more than inherent variation.

Step 2. Fill in background information on the p-chart form.

Figure 6-7 is the form for p-charts. This form makes it easy to write down all the necessary background information, such as form/unit number and operation/task. It also has a graph for plotting points.

Fill in the background information.

Step 3. Collect samples and record data.

Your sample is 50 forms. We suggest you take at least 20 samples.

You have learned from the office manager how to tell whether or not a form is defective, good or bad. You have a list of the various types of errors that includes various defects such as missing information, typos, math errors, misspellings, wrong box, and other. Now check each form and write down the following information under the correct headings. (1) Record the date and the time if that's important. (2) Under "NO. INSP." (number inspected), write 50. This is your sample size. (3) Under "NO. DEF." (number of defectives), write 18, the number of defective units you found in the first sample. (4) Under "KINDS OF DEFECTS," write down headings for the errors from the list such as "typos" and "math errors." Count the specific errors. Remember, this total count of errors may be larger than the number of defective forms because a defective form may have more than one error. In this first sample, the number of defective

Figure 6-7. p-chart with background information and first sample recorded.

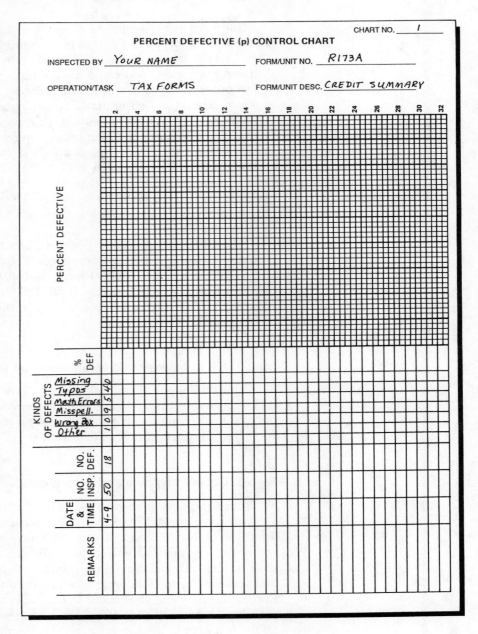

forms is 18, but the number of errors is 19 because one form had two kinds of errors, a math error and a misspelling. See Figure 6-7.

Step 4. Calculate p, percent defective.

You have inspected the 50 forms and recorded all the data on the p-chart. The next thing to do is to calculate the p, the percent defective. Remember, the formula for p is

p equals the number of defectives divided by the sample size times 100%

For the first sample: $p = (18 \div 50) \times 100\% = 36\%$.
For the second sample: $p = (18 \div 50) \times 100\% = 36\%$. See Figure 6-8.

Step 5. Calculate the average percent defective for the process.

When you have finished taking your 20 samples, add up all the numbers under "NO. DEF."

Number of defectives:

18, 18, 12, 17, 17, 12, 9, 13, 14, 11,

23, 6, 14, 8, 8, 13, 18, 14, 14, 17

Total number of defectives: 276.

Divide the total number of defectives by the total number of forms you checked. Then multiply by 100%. You have looked at 50 forms 20 times, so you have inspected a total of 1000 forms.

Average percent defective for the process, $\bar{p}$:

$(276 \div 1000)$ times 100% equals 27.6%

Don't enter the $\bar{p}$ on your chart yet. Make a note of it for the next step.
Check your arithmetic.

Step 6. Determine the scales for the graph and plot the data.

Set the scale for the graph so that the largest and smallest values for p can fit easily inside the ends of the scales. (Often the lower control limit for p is zero.) Be sure to leave enough room above the largest p for the upper control limit. Pick scales that make it easy to plot the data and read the chart.

Figure 6-8. p-chart with p calculated for each sample.

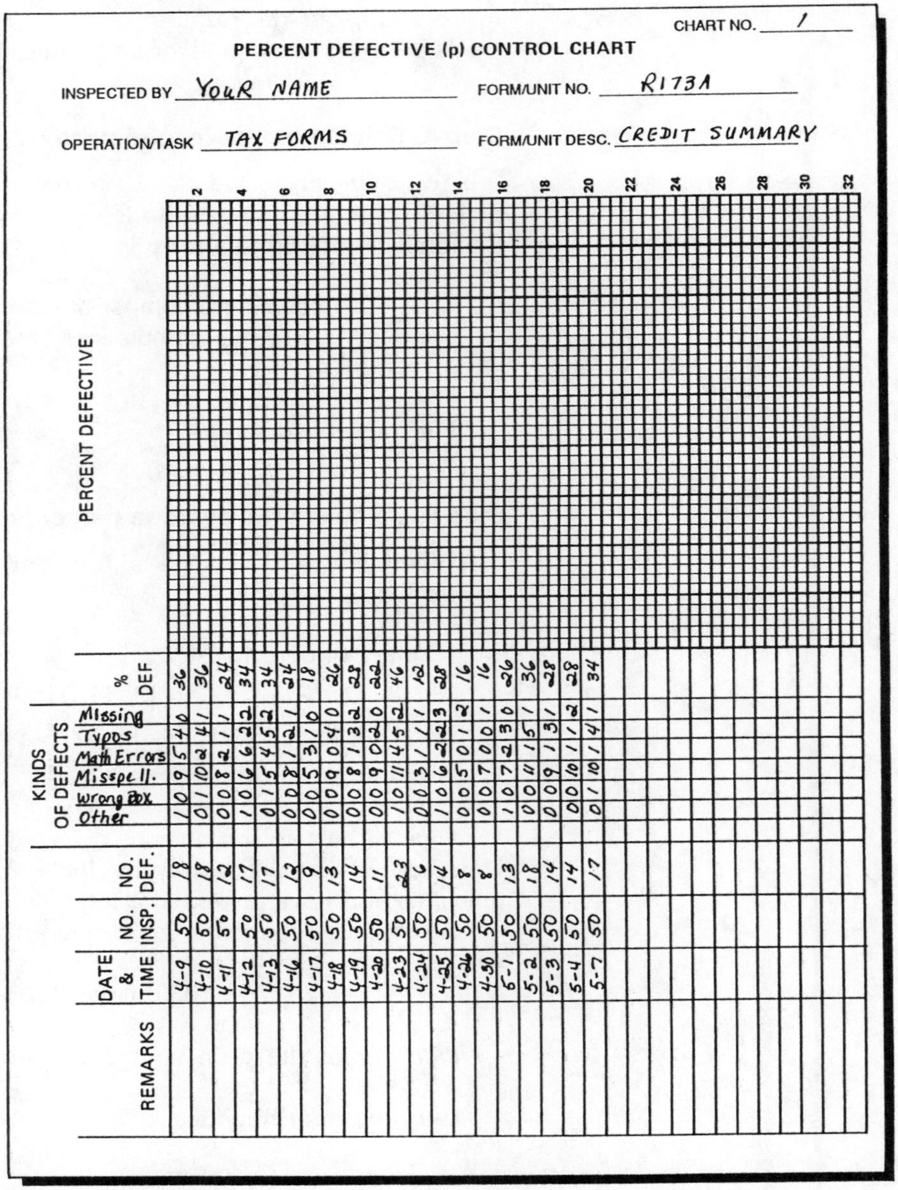

PERCENT DEFECTIVE (p) CONTROL CHART

CHART NO. 1
INSPECTED BY: YOUR NAME
FORM/UNIT NO. R173A
OPERATION/TASK: TAX FORMS
FORM/UNIT DESC. CREDIT SUMMARY

% DEF	36	36	24	34	34	18	26	28	46	12	28	16	26	36	28	34
Missing	6	4	1	2	3	3	4	1	13	0	4	1	2	5	1	0
Typos	4	4	1	6	5	1	4	5	2	1	2	2	3	1	1	4
Math Errors	9	8	7	6	4	3	5	3	9	0	4	3	5	7	5	1
Misspell.	0	1	2	1	5	2	0	3	0	0	1	1	1	3	5	1
Wrong Box	0	0	0	0	0	0	0	0	0	1	0	0	1	0	0	0
Other	0	1	1	1	0	0	0	1	0	0	3	0	0	2	1	1
NO. DEF.	18	18	12	17	17	12	9	13	14	11	23	6	14	8	8	13	18	14	17
NO. INSP.	50	60	50	50	50	50	50	50	50	50	50	50	50	50	50	50	50	50	50
DATE & TIME	4-9	4-10	4-11	4-12	4-13	4-16	4-17	4-18	4-19	4-20	4-23	4-24	4-25	4-26	4-30	5-1	5-2	5-3	5-4	5-7

Largest percent defective (p): 46
Smallest percent defective (p): 12

You can run your scales from zero to 60. All the p's will fit easily and still leave plenty of room for the upper control limit. On your graph,

Figure 6-9. p-chart with data plotted and $\bar{p}$ drawn in.

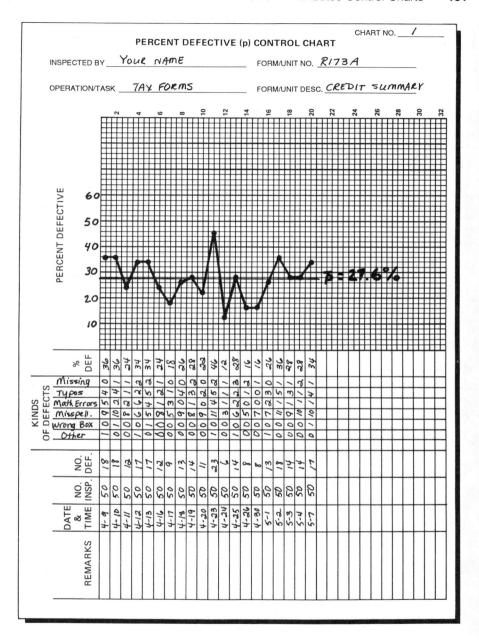

there will be five lines between 10 and 20, so each line represents 2.0 units. It will be easy for you to plot the p's. See Figure 6-9.

Now plot the p's and connect the plot points with straight lines.

Finally, draw a solid line across the graph, at 27.6% for your average percent defective, $\bar{p}$, and label it "$\bar{p} = 27.6\%$."

Step 7. Calculate the control limits for percent defective charts.

Calculating control limits for p-charts is not quite as easy as calculating the limits for average and range charts. There is less work to do, but it is a little more difficult.

Here is the formula for calculating UCL_p, the upper control limit for p:

$$UCL_p = \bar{p} \text{ plus } 3 \sqrt{\frac{\bar{p} \times (100\% - \bar{p})}{n}}$$

- $\sqrt{}$ is the symbol for the *square root*, a special calculation that mathematicians have worked out.
- × is the symbol for "times."
- n is the sample size.
- $\bar{p}$ is the average percent defective.

This may look like a lot of "chicken tracks," but it's really not too tricky. Read it this way: The upper control limit for p equals $\bar{p}$ (the average percent defective) plus 3 times the calculation from the square root symbol.

Inside the square root symbol, first calculate 100% minus $\bar{p}$. Then multiply the result by $\bar{p}$. Divide that result by n, the sample size. Once you have calculated all this, take the square root by entering the number on a calculator and pressing the key marked $\sqrt{}$.

Multiply this result by 3 and then add it to $\bar{p}$, the average percent defective.

The formula for LCL_p, the lower control limit for p, is

$$LCL_p = \bar{p} \text{ minus } 3 \sqrt{\frac{\bar{p} \times (100\% - \bar{p})}{n}}$$

Do all the same calculations as for the upper control limit, but this time subtract 3 times the number you obtained from the square root symbol. If you obtain a negative number when you figure the lower control limit, set the lower control limit at zero.

This is how your calculations will look when $\bar{p}$ equals 27.6% and the sample size (n) is 50.

Upper control limit (UCL_p):

$$UCL_p = \bar{p} + 3 \sqrt{\frac{\bar{p} \times (100\% - \bar{p})}{n}}$$

$$= 27.6\% + 3 \sqrt{\frac{27.6\% \times (100\% - 27.6\%)}{50}}$$

$$= 27.6 + 3 \sqrt{\frac{27.6 \times 72.4}{50}}$$
$$= 27.6 + 3 \sqrt{\frac{1998.24}{50}}$$
$$= 27.6 + 3\sqrt{39.9648}$$
$$= 27.6 + 3 \times 6.32$$
$$= 27.6 + 18.96$$
$$= 46.56\%$$

Lower control limit (LCL_p):

$$LCL_p = \bar{p} - 3 \sqrt{\frac{\bar{p} \times (100\% - \bar{p})}{n}}$$
$$= 27.6\% - 18.96\%$$
$$= 8.64\%$$

If your answer were negative, you would set the lower control limit at zero.

Be sure to check all the steps and all your arithmetic.

Now draw the control limits on your p-chart and label them UCL_p and LCL_p. Make a note of this step under "REMARKS." We suggest that you draw the limits as dashed lines or in color so that they are clearly visible. See Figure 6-10.

Fraction Defective p-Charts.

Sometimes people set up another kind of p-chart that uses fractions instead of percentages, called *fraction defective p-charts* or *proportion defective p-charts*. This kind of p-chart shows how many units are defective compared to the total in the sample. Figure 6-11 shows the chart for Credit Summary Form R173A data using fractions.

The fraction defective charts look just about the same as the percent defective charts, but there are three differences. See Figures 6-10 and 6-11.

The first difference is in the calculation of p, the fraction defective. The formula for p using fractions, not percentages, is

p equals number of defective units divided by sample size.

For Form R173A, the first p would be

$$p = 18 \div 50$$
$$= 0.36$$

Figure 6-10. p-chart with upper and lower control limits set.

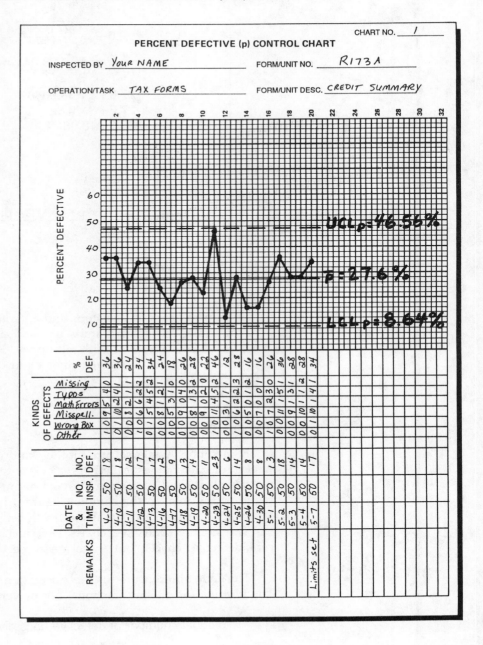

The next p would be

$$p = 18 \div 50$$
$$= 0.36$$

Figure 6-11. Fraction defective p-chart.

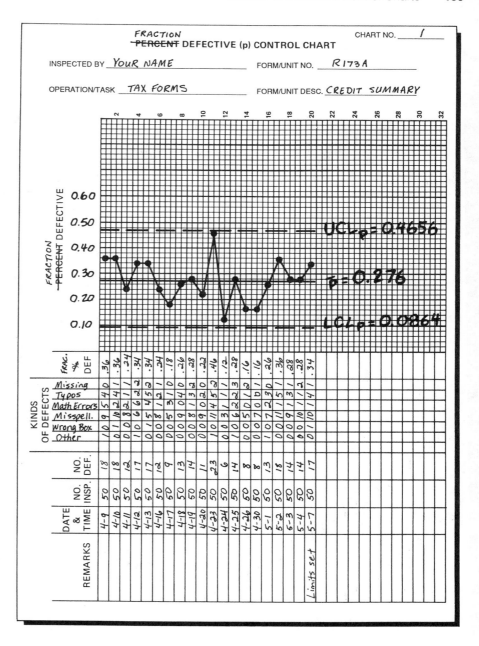

Do the remaining calculations for p in the same way. When you calculate p, you are *not* multiplying by 100%, as you did in the percent defective chart. You get the same answer, but the decimal point is two places to the left. Calculate $\bar{p}$ as follows: Add up all the defectives and then divide by the total number of units inspected. In the credit summary example, the number for $\bar{p}$ is the same, but the decimal point is two places to the left:

$$\bar{p} = 276 \div 1000 = 0.276$$

(Note that this is not a percentage, but a decimal fraction.)

The second difference is in how you calculate the upper and lower control limits. The formula for the upper limit for p, fraction defective, is

$$UCL_p = \bar{p} + 3\sqrt{\frac{\bar{p} \times (1 - \bar{p})}{n}}$$

The numbers inside the square root symbol are different. You are using 1 instead of 100%, and $\bar{p}$ is a decimal fraction, not a percentage. Otherwise you make the same calculations as in the percent defective chart.

This is how all the "chicken tracks" look when you work out the UCL_p for the credit summary forms using fraction defective:

$$\begin{aligned} UCL_p &= 0.276 + 3\sqrt{\frac{0.276 \times (1 - 0.276)}{50}} \\ &= 0.276 + 3\sqrt{\frac{0.276 \times 0.724}{50}} \\ &= 0.276 + 3\sqrt{\frac{0.199824}{50}} \\ &= 0.276 + 3\sqrt{0.00399648} \\ &= 0.276 + 3 \times 0.0632 \\ &= 0.276 + 0.1896 \\ &= 0.4656 \end{aligned}$$

You see that you have the same answer except for the decimal point, which is two places to the left.

Calculate the lower control limit with this formula:

$$\begin{aligned} LCL_p &= \bar{p} \text{ minus } 3\sqrt{\frac{\bar{p} \times (1 - \bar{p})}{n}} \\ &= 0.276 - 0.1896 \\ &= 0.0864 \end{aligned}$$

You should set your lower control limit at zero if the final calculation is negative.

The third and last difference is that the scale for the fraction defective p-chart is shown in decimals, not percentages. Your scale for the p-chart in this example will read 0.10, 0.20, 0.30, and so on. Be sure to cross out "PERCENT" on the chart and write in "FRACTION."

We are telling you about the fraction defective p-chart only because you may run into it once in a while. Most people prefer to do their calculations for p-charts in percentages, not in fractions. It's easier to think in terms of percentages—8% defective or 12% defective—than in terms of decimals—0.08 defective or 0.12 defective.

Step 8. Interpret the p-chart.

When you are interpreting the p-chart you have just set up, there are three possible situations you need to consider. Either *all* the percents defective, p's, are *inside* the control limits; or there are *one or two* p's *outside* the limits; or *three or more* p's are *outside* the limits. (You have already seen this kind of decision making in Module 5.) Remember, you may use this three-step rule *only* when you are first setting up a new p-chart.

If *all* the p's are within the control limits, the process is in statistical control. Only inherent variation is at work, as far as you can tell.

If *one or two* p's are outside the limits, the usual practice is to throw out those one or two p's and the samples from which they come. Completely refigure the average percent defective, $\bar{p}$. Then, using the new $\bar{p}$, calculate new control limits.

Once you have recalculated the control limits, there are two situations to check. If one or more of the remaining p's are still outside the new control limits, your process is not in statistical control. Assignable causes are disrupting the process. Find and remove them. Then take new samples and data and set up the p-chart again, using limits based on this new information.

On the other hand, if all the p's are now inside the new control limits, go ahead and run your process. But be careful! Some assignable causes may still be around to cause trouble later.

If *three or more* p's on the initial chart are outside the limits, your process is clearly out of statistical control. First, you must clean out the assignable cause or causes that are upsetting your process. Then you can start again. Collect new data and set up new p-charts.

Figure 6-12 is a decision chart to help you understand how to interpret p-charts. It is similar to the decision charts you saw in Module 5.

Figure 6-10 shows the p's for Credit Summary Form R173A. Every p lies between the upper control limit of 46.56% and the lower control limit of 8.64%. Therefore, the p's are in statistical control.

HOW TO USE A NEWLY DEVELOPED p-CHART IN CONTINUED OPERATIONS

Once the percentages of defectives are in control, you can say that your process is in statistical control. That is, it appears that only inherent

Figure 6-12. Decision chart for working with p's.

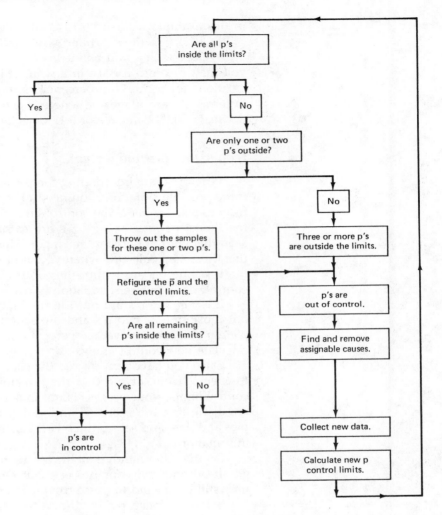

variation is at work and is present because of chance causes. Apparently, no assignable causes are present.

A word of warning: Even though the process is in control, it may not be doing what you want. The level of the $\bar{p}$ may not be acceptable because of cost or customer dissatisfaction. You and your management will have to work that out.

Now that the p-chart is set up and the percents defective are in control, you can use the chart in an ongoing manner. Once you have determined your control limits, you can use them to tell whether errors are due to chance causes or to assignable causes.

Whenever a single point falls outside the control limits, it is evidence that the process is out of statistical control. If the point is out on the high side above the UCL$_p$, find the assignable cause and remove it. If the point is on the low side below the LCL$_p$, look to see what assignable cause is improving the process and duplicate it!

Review your p-chart periodically to make sure you're still operating where you think you are. You will need to review the p-chart if you have tried to make some improvements, or if you think something is different about the process. Take 10 new samples of 50 forms as close together as possible. Find the new p values and calculate the new average ($\bar{p}$). Now you can compare the old $\bar{p}$ with the new one to see how the process is doing.

THE np-CHART

Sometimes you will find it convenient to use a control chart that plots the *number* instead of the percentage of defective units in the sample. This kind of chart is called an np-chart.

One rule must be followed strictly when you use an np-chart: The size of every sample must be the same.

SETTING UP THE np-CHART

Setting up the np-chart is very similar to setting up a p-chart. The only difference is that the number of defective units is recorded instead of the percent defective or fraction defective.

Step 1. Take the sample.

The sample size for the np-chart should be large enough to make sure that you will see defective units in each sample. A good rule to follow is to set your sample size so that the average number of defective units per sample will be 4 or more. Take your sample in the same way and with the same frequency as you would for a p-chart.

Step 2. Fill in the form for data and graph.

You can use the same chart form for the np-chart as you use for the p-chart. Simply cross out the words "PERCENT DEFECTIVE" and write in "NUMBER DEFECTIVE (np)" to identify the chart as an np-chart. Nothing will be written under "% DEF.," so cross it out.

Step 3. Collect samples and record data.

This step is identical to the p-chart procedure. "NO. INSP." will have the same number for each sample. Remember, you plot the number of defective units in each sample, not the percentage or fraction of defective units on the chart.

Step 4. Calculate the n$\bar{p}$.

To calculate the n$\bar{p}$, or the *average number defective for the process*, add up the total number of defective units found in the samples and divide that total by the number of samples. Remember that in the np-chart, if you look at (say) 50 units at a time, those 50 units make up a single sample. The total number of samples is the total number of times you looked at units. It is not the total number of units inspected.

The formula for finding the n$\bar{p}$ is

> n$\bar{p}$ equals the number of defective units in all samples divided by the number of samples taken.

The average number defectives for the process is n$\bar{p}$.

Let's use the inspection data shown in Figure 6-8 to make an np-chart. The number of defectives has been transferred to Figure 6-13, along with the other information from Figure 6-8.

To calculate the process average number defective for our example, add all the numbers under "NO. DEF.," which total 276. Then divide 276 by the number of samples, which is 20. The result is 13.8, the n$\bar{p}$. Draw a solid line on the chart at the 13.8 level and write in "n$\bar{p}$ = 13.8."

Step 5: Determine the scale for the graph and plot the np's.

Set the scale so that the largest and smallest np's will fit on the graph, and pick a scale that makes it easy to plot the data. Be sure to leave extra space for the upper control limit. (The lower control limit will often be zero.)

Step 6. Calculate the control limits for number defective.

The formula for calculating the control limits for np, number defective, is similar to the formula for finding the control limits for p, percent defective, on the p-chart. This is how your calculations will look when n$\bar{p}$ equals 13.8 (n$\bar{p}$ is the average number defective and n is the sample size):

Upper control limit for np:

$$UCL_{np} = n\bar{p} + 3\sqrt{n\bar{p} \times (1 - n\bar{p}/n)}$$
$$UCL_{np} = 13.8 + 3\sqrt{13.8 \times (1 - 13.8/50)}$$
$$= 13.8 + 3\sqrt{13.8 \times (1 - 0.276)}$$
$$= 13.8 + 3\sqrt{13.8 \times 0.724}$$

(equation continues)

Figure 6-13. np-chart.

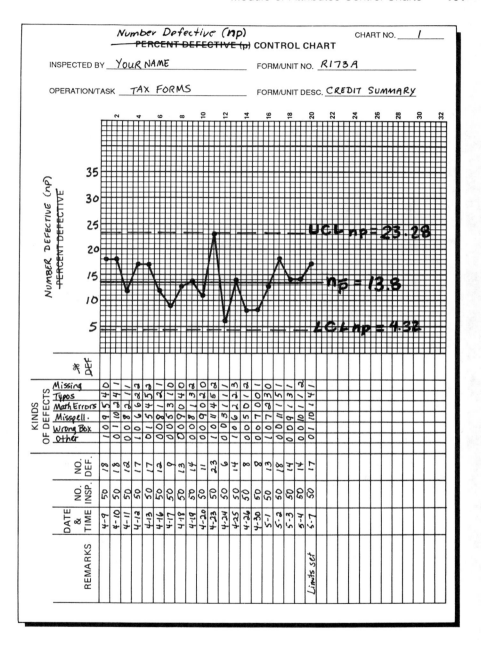

$$= 13.8 + 3\sqrt{9.9912}$$
$$= 13.8 + 3(3.1609)$$
$$= 13.8 + 9.4827$$
$$\text{UCL}_{np} = 23.2827$$

Draw the upper control limit on the chart as a dashed line. See Figure 6-13.

To find the lower control limit for number defective, use this formula:

$$LCL_{np} = n\bar{p} - 3\sqrt{n\bar{p} \times (1 - n\bar{p}/n)}$$
$$= 13.8 - 9.4827$$
$$LCL_{np} = 4.32$$

(If this answer were a negative number, you would set the lower control limit at zero.)

Step 7. Interpret the np-chart.

Interpret the newly developed chart in the same way you did the p-chart. (See Step 8, "Interpret the p-chart.")

This control chart looks very much like the p-chart in Figure 6-10. The difference is that the p-chart shows the *percent* defective in the sample and the np-chart (Figure 6-13) shows the *number* defective in the sample. You may choose to use either chart, but each sample for the np-chart must be the same size.

c-CHARTS

Earlier in this module, we told you that you can control a nonmeasurable characteristic in your service by using a p-chart, an np-chart, or a c-chart. The p-chart helps you control the percentage of defective units in the operation and the np-chart helps you to control the number of defective units, but the c-chart helps you control the number of defects in a *single element or group of elements* that make up the service. It's a good idea to use a c-chart when there are opportunities for one or more defects to be present in a unit or a group of units. Figure 6-14 is a c-chart.

The c-chart should not be confused with the np-chart. As you remember, the np-chart plots the *number of defective units* in the sample. The c-chart plots the *number of defects* in the *inspection unit*.

As you know, the p-, np-, and c-charts all use attribute or counting data. For the p-chart and np-chart you counted how many units in your sample were defective, that is, not good. For the c-chart, you will be counting how many specific defects there are in each inspection unit. An inspection unit may consist of one element, such as a waxed floor; or it may consist of a group of units, such as a patient's breakfast tray with its cereal, coffee, eggs and toast. One unit or group of units may have any number of defects, or it may have none at all.

It is very easy to calculate c: just count up all the *defects* in each inspection unit. There is no formula.

Figure 6-14. c-chart already set up. (Form developed by Robert and Davida Amsden and Howard Butler.)

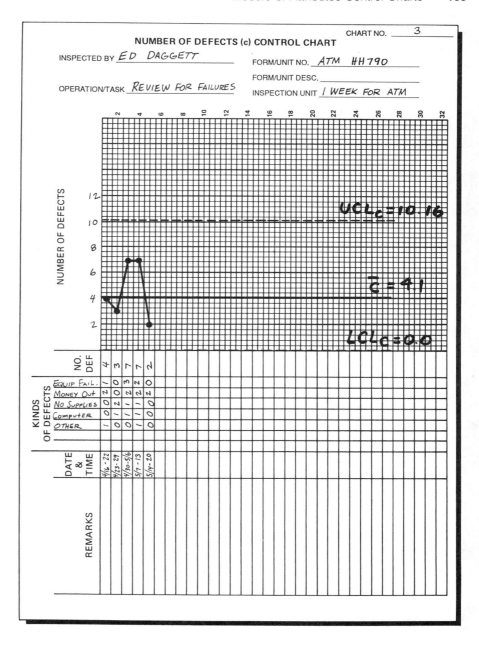

HOW TO USE c-CHARTS

Earlier in this module, we looked at a p-chart that somebody else had already set up. Now we will do the same thing with c-charts.

As with the other attribute charts, your job is to take the sample, count the number of defects or errors in the inspection unit, write that data on the c-chart, and plot the point. You also have to understand what the chart is saying so that you'll know whether to keep doing the job or to make corrections.

The c-chart in Figure 6-14 is monitoring the number of failures of an automatic teller machine, or ATM. Your job is to review ATM transactions for the week and count the total number of failures and how many of each kind of failure there are. Somebody else set up the chart originally and marked the first few samples for you.

You will be marking the chart for this week. On Monday, review the transactions from the previous week and count the failures. Write your name above Ed's on the chart and fill in the time when you started marking the chart. See Figure 6-15.

On the chart are several headings under "KINDS OF DEFECTS." For this example, think of "defect" as a failure because failures are what you are tracking. Each heading has room for one of the particular kinds of failures that you are looking for and are counting: "equipment failure," "money out," "no supplies," "computer failure," and "other." You know that an ATM may have more than one kind of failure: the machine may be out of money or supplies; there may be a failure in the on-line computer. Simply count up how many of each failure you find and write that number in the proper place.

Now add up all the kinds of failures you have found and write this number under "NO. DEF." (number of failures). There are one "money out," two "no supplies," one "computer failure," and nothing else. The sum of these defects is 4. Write this number, 4, under "NO. DEF." When you have filled in the data you can plot the point as a large dot on the graph. Connect this dot with the previous dot by drawing a straight line. See Figure 6-15.

INTERPRETING c-CHARTS

Just as averages, ranges, and percents defective have upper and lower control limits, so do c's (failure counts). Remember this when you check your plot.

Ask yourself, "Is the c where it's supposed to be?" In this case, the c is 4. It's between the upper control limit of 10.16 and the lower control limit of zero, so it's O.K.

What does the chart tell you? When your c (failure count) is within the control limits, your process is in statistical control. This means you can continue to use this ATM without making any corrections or adjustments to the process.

Everything we said about points inside the control limits, near the limits, or outside the limits, as well as patterns of points, applies as much to c-charts as to p-charts. (See this discussion in the p-chart section of this module.) You want to find out whether your process is in control, that

Figure 6-15. c-chart with new data and new point plotted.

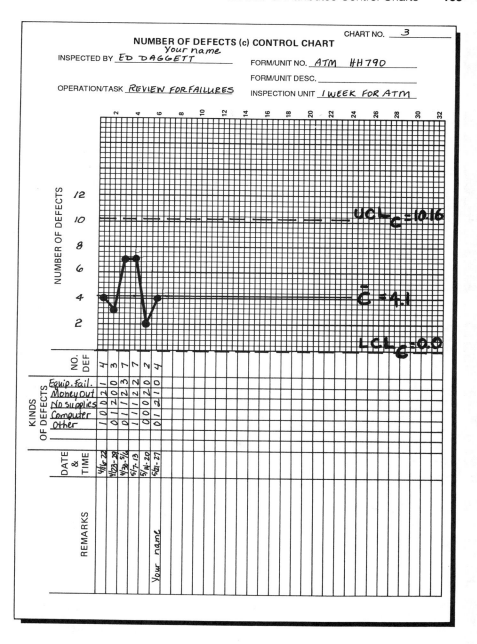

is, whether only inherent variation due to chance causes is present or whether there are any assignable causes of variation. The c-chart provides you with this kind of information.

If the c-chart shows a point or points outside the limits, you must hunt down the assignable causes and remove them. You have already learned the sources of assignable causes: equipment, method, materials, environment, or people. We talked about these sources in Modules 2 and 5 and the p-chart section of this module.

SETTING UP c-CHARTS

Suppose you go to a different bank branch to monitor the failure rate of an automatic teller machine. You are responsible for setting up the c-chart. A c-chart for this process will do several things. It will help you monitor the process and keep track of the number and kinds of failures. It will also document what is happening and communicate all this to you, your supervisor, the bank's quality assurance department, and anyone else who wants to know.

Step 1. Take the sample.

You want your sample to show only inherent variation. To do this, the sample must represent only one source of data. Therefore, you will take the sample from one piece of equipment, one employee, one time period.

The *inspection unit* is all the transactions for one ATM for one week (see Figures 6-16 and 6-17). For a hospital, the inspection unit might be all the accidents to staff for one month. For a school, the inspection unit might be the number of incidents of breakage in the cafeteria per week.

Remember that once you establish the inspection unit, you must stick to it. If the inspection unit is the output of one ATM for one week, always use *one* week's transactions from the ATM as your sample. If you decide to use one month's transactions as the sample unit, then always review a month's transactions.

Step 2. Fill in background information on the c-chart form.

The layout of the c-chart is similar to the p-chart. See Figure 6-16. The c-chart has a place for information such as operation/task, form/unit number, and inspection unit. Fill in this background information.

Step 3. Collect and record data.

Follow the plan you established and review one week's batch of the ATM's transactions.
Follow the standards for determining failures.
Record the data on the chart.
The supervisor has shown you what failures to look for. When you review each week's transactions, count how many of each failure you

Figure 6-16. c-chart with background information and data recorded.

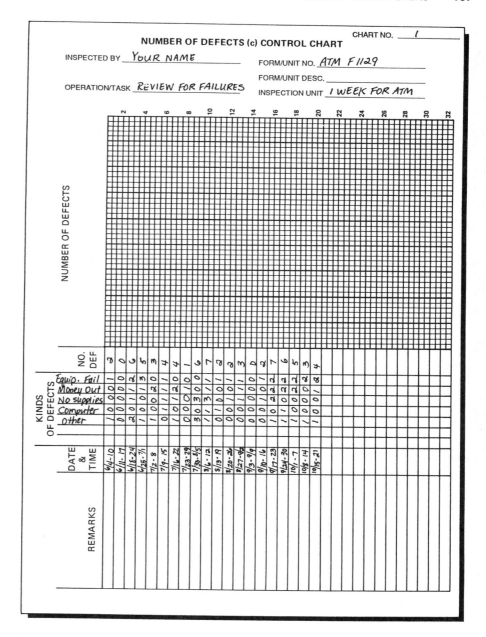

find. Write these numbers in the appropriate places under the heading "KINDS OF DEFECTS." Add up the total number of defects for the first sample and write that number, 2, under "NO. DEF." This total, 2, is what we call "c," or count. Note down anything of interest under "REMARKS."

Continue in this way until you have recorded data for 20 samples. See Figure 6-16.

Step 4. Calculate the average number of defects for the process.

Add all the numbers of defects, c's, under "NO. DEF."
Divide the total of the c's by the number of inspection units. In this case, an inspection unit is the week's transactions for ATM F1129.
Check the arithmetic.

Defects, c: 2, 0, 6, 5, 3, 4, 4, 1, 6, 7, 2, 2, 3, 0, 2, 7, 6, 5, 3, 4

Total of c's: 72

Number of c's: 20

Now calculate *the average number of defects for the process*, $\bar{c}$:

$$\bar{c} \text{ equals total of all defects (c's) divided by number of inspection units}$$
$$\bar{c} = 72 \div 20$$
$$= 3.6$$

Make a note of the $\bar{c}$. You will add it to the chart later.

Step 5. Determine the scales for the graph and plot the data.

Find the largest and smallest c. The largest c is 7 and the smallest c is 0.
Set scales on the graph so the largest and smallest c's can fit easily. Allow enough room for the control limits.
You can run the scales from zero to 10. All the c's will fit, and you'll have room for your upper and lower control limits. On the graph, the first dark line will be 2, the next dark line 4, and so on.
Plot the c's and connect the points with straight lines. Now draw in the overall mean, $\bar{c}$, and label it 3.6. See Figure 6-17.

Step 6. Calculate the upper and lower control limits.

The arithmetic for calculating the control limits for c-charts is easier than for p-charts. This is the formula for the upper control limit for c:

$$UCL_c = \bar{c} + 3\sqrt{\bar{c}}$$

The upper control limit for c equals $\bar{c}$ plus three times the square root of $\bar{c}$. (To do square roots easily, use a calculator.)

Figure 6-17. c-chart with points plotted, c̄ drawn in, and control limits set.

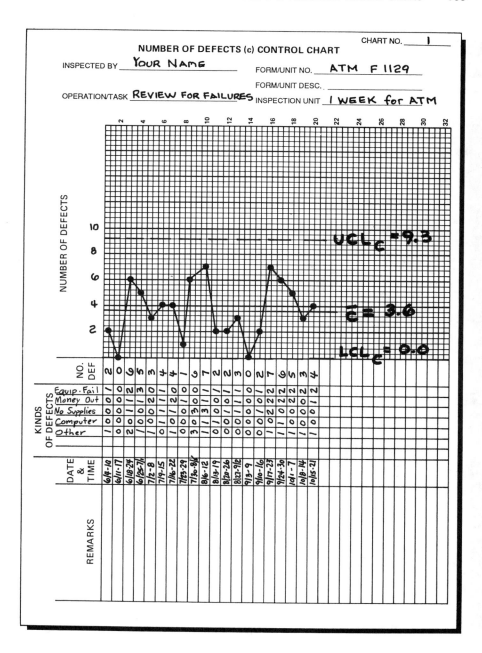

To figure the lower control limit, use this formula:

$$LCL_c = \bar{c} - 3\sqrt{\bar{c}}$$

The lower control limit for c equals $\bar{c}$ minus three times the square root of $\bar{c}$.

Remember, when your figure the lower control limit, you are *subtracting* from $\bar{c}$ the term three times the square root of $\bar{c}$. If your answer is a negative number, set it at zero.

This is how your calculations will look when $\bar{c}$ equals 3.6.

Upper control limit for c:

$$\begin{aligned} UCL_c &= \bar{c} + 3\sqrt{\bar{c}} \\ &= 3.6 + 3\sqrt{3.6} \\ &= 3.6 + (3 \times 1.897) \\ &= 3.6 + 5.691 \\ &= 9.291 \end{aligned}$$

Round off 9.291 to 9.3.

Lower control limit for c:

$$\begin{aligned} LCL_c &= \bar{c} - 3\sqrt{\bar{c}} \\ &= 3.6 - 3\sqrt{3.6} \\ &= 3.6 - (3 \times 1.897) \\ &= 3.6 - 5.691 \\ &= -2.091 \end{aligned}$$

The answer is negative, so set the lower limit at zero.

Check your arithmetic at every step. You don't want to be an assignable cause!

Draw in the control limits on your c-chart and label them UCL_c, 9.3 and LCL_c, 0.0. Show these as dashed lines or use color so that they are highly visible. See Figure 6-17.

Step 7. Interpret the c-chart.

Once you have completed the c-chart, there are three possible situations, just as with the p-charts. Either all the c's are inside the control limits; or one or two c's are outside the control limits; or three or more c's are outside the control limits. As you did for p-charts, take the appropriate action, depending on the number of c's outside the control limits.

As you can see in Figure 6-17, all the c's for ATM F1129 transactions are in control. Because all the c's are in control, you can use your c-chart to continue monitoring and controlling your process.

SUMMARY

When you need to control or monitor quality characteristics in the process of producing or delivering a service, you may use an attributes control chart. These charts are based on attributes or counting data. An attribute is not measured; it's either there or it is not.

The three most important kinds of attributes charts are the p-chart, the np-chart, and the c-chart.

Use the p-chart when you want to control the percent or fraction defective of the units in the service. If you want to monitor the number of defectives, use the np-chart. To control the number or count of defects in the inspection unit, use the c-chart.

A unit or item is defective if it is not good (not acceptable). A defective unit may have one or more defects, such as a typing error or missing information, which makes the unit unacceptable or defective.

You must know what your control charts are telling you. Points inside the control limits mean that as far as anybody can tell, only inherent variation is affecting your process. Therefore, no corrections or adjustments are necessary. If one or more points lie outside the limits, however, or if certain patterns of points appear, then assignable causes are disrupting your process. At this point, some adjustment or correction will be required.

A point outside the control limits tells you that something about the operation has changed. It does not tell you *what* happened, only *when* it happened.

PRACTICE PROBLEMS: ATTRIBUTES CHARTS

Work through the following practice problems using the statistical techniques you learned in Module 6. The solutions to these problems can be found in the "Solutions" section beginning on page 266.

Problem 6-1.

You work for a home improvement center. For a long time, customers have been complaining about dirt and grease spots on the wall panels they pick up at the storage/loading area. Something must be done to improve this situation.

As panels were being loaded for customers, someone inspected 48 panels each day for 20 days.

(a) Using this data, try to determine whether the problem is coming from assignable causes or from system causes.

(b) Can this problem be solved by the people in the loading area, or does management need to do something to bring about improvement?

Day	No. Defectives	Day	No. Defectives
1	30	11	25
2	27	12	20
3	34	13	24
4	23	14	23
5	20	15	16
6	33	16	19
7	22	17	32
8	24	18	26
9	18	19	25
10	34	20	25

Problem 6-2.

A toll collection agency in a major urban area records all transactions by the toll collectors. For each transaction, management records the number of axles passing through the toll booth as well as the amount of money collected. Then they select 2 days per month and check these records to see how many times the number of axles does not properly match the amount of money collected. One axle should show an associated transaction of 25 cents, two axles should show 50 cents, and so forth. The data represent the number of errors for the 2-day period.

(a) Are the three toll collectors in control?

(b) Do you see any differences between the averages for the three toll collectors?

(c) Management is aware that when a new toll collector starts work, that collector experiences relatively high error rates that decrease and eventually settle down to a stable average. Do you see any evidence of this in your charts? If so, how many months do you think it takes to level off?

(d) Management has stated that an error rate of six errors per 2 days of transactions is a good rate. The three data sets represent the results from the very best collector, someone in the middle, and one of the worst. Do you feel that management's specification of six errors in 2 days is a reasonable figure?

\multicolumn{3}{c	}{Error Rates}	
\multicolumn{3}{c	}{Toll Collector}	Month
A	B	C	
12	58	30	Oct
14	47	29	Nov
4	24	15	Dec
7	32	215	Jan
10	39	22	Feb
6	23	17	Mar
5	17	10	Apr
9	24	11	May
6	34	14	Jun
3	27	9	Jul
4	12	21	Aug
7	21	14	Sep
7	18	—	Oct
13	17	24	Nov
6	20	15	Dec
7	30	17	Jan
4	15	19	Feb
6	18	12	Mar
9	19	8	Apr
3	16	9	May
7	14	9	Jun
5	23	11	Jul
3	27	9	Aug
3	21	12	Sep

MODULE 7

Sampling Plans

NEW TERMS IN MODULE 7
(in order of appearance)

acceptance sampling	sampling biases
producer's risk	random sample
consumer's risk	inspection lot
Military Standard (Mil-Std) 105E	operating characteristic curve
sample size	OC curve
n	acceptable quality level
acceptance number	AQL
c	lot tolerance percent defective
lot size	LTPD
k	average outgoing quality
probability of acceptance	AOQ
P_a	average outgoing quality limit
representative sample	AOQL

INTRODUCTION

Sampling plans, the subject of this module, may not appear to be a statistical process control technique like the control charts discussed in the other modules. However, just like the problem-solving tools in Module 2 and the quality improvement tools in Module 3, sampling plans are tools of quality to be used in conjunction with the SPC tools of quality.

When reading this module you may decide that designing or selecting a sampling plan is not part of your job, as would be, say, the development and use of a control chart. If that is the case, you may want to concentrate on the other modules of this book. In this module we will discuss what sampling plans are, how they are used, where to find tables of standard sampling plans, and how to develop your own sampling plans to suit your particular needs.

Services are produced and delivered just as manufactured goods are produced and delivered. Services have characteristics, which are created when the services are produced. These characteristics may appear on order forms, on bank statements, or in any number of places in the process of delivering a service to a customer. They may even be a part of the service itself, such as the condition of a product upon delivery. The absence or the inaccurate presence of these characteristics often will affect the quality of the service delivered.

Many of the characteristics or items involved in the process of delivering a service are produced in large quantities. The individual units must be without defect if the service offered to the customer is to be fully satisfactory. A defective individual item may not prevent delivery of the service to the customer, but it may have to be corrected or replaced for the customer.

The process used to produce these items may not be capable of producing all the items without defects all the time. In many instances, having such items 100% defect free may not be critical to the delivery of the service, but only so long as the rate of occurrence of the defects is maintained at some low level. If the level of defects gets too high, you will lose customers. On the other hand, complete elimination of defects *by sorting* may not be possible, and developing a completely new and different process to eliminate all defects may not be economically practical.

In these cases, management must decide on a defect level the company can live with for the time being and still be successful in meeting the wants, needs, and expectations of the customer. Once this has been decided, it is necessary to assure that the specified level of quality has been achieved and maintained. One way to accomplish this is through the use of acceptance sampling plans.

ACCEPTANCE SAMPLING

Acceptance sampling means inspecting a sample of a stated number of items from a group, or lot, of those items and using this sample to judge the quality of the entire lot. If the quality of the sample is acceptable, you accept the whole lot. If the quality of the sample is unacceptable, you reject the lot. Rejected lots are inspected 100%, sorted, and corrected, or they may in some instances be scrapped.

Let's look at a simplified example of acceptance sampling in action. At the Big Party Planner's Company shipments of balloons are received periodically at the company warehouse. There are 75 balloons in each box. Employees want to make sure the balloons are in good condition before they deliver them to party locations—but they can't test every balloon before it's sent out. To solve this problem, Joe, the warehouse supervisor, selects 13 balloons at random from each box of balloons received and inspects them. If 1 or more of the balloons in the sample has a defect,

Joe rejects the box of balloons and has someone inspect and sort every balloon in the box. If the sample doesn't have any defects, Joe accepts the box and lets the balloons be shipped.

How did Joe determine the number of balloons to be taken from each box for inspection? He could arrive at the number of balloons to be inspected, called the sample size, in several ways. He could merely pick a number and call that the sample size, or he could be a little more scientific and refer to one of several published tables of sampling plans, or he could design his own sampling plan. Regardless of which method Joe used, he must also determine the number of defective balloons he will accept in the sample. This is called the acceptance number. A third thing Joe must do in order to establish a sampling plan is set the overall level of quality he wants to accept. This is called the acceptable quality level.

In this case, Joe was willing to accept 1% defective, on the average, as an acceptable quality level. With this in mind along with the knowledge that the balloons came 75 to a box, Joe referred to two tables of standard sampling plans called Military Standard 105E—Sampling Procedures and Tables for Inspection by Attributes. The first table Joe used is one for Sample Size Code Letters. (See Table 7-1.) The lot size Joe used is

TABLE 7-1
Sample size code letters.

Lot or batch size			Special inspection levels				General inspection levels		
			S-1	S-2	S-3	S-4	I	II*	III
2	to	8	A	A	A	A	A	A	B
9	to	15	A	A	A	A	A	B	C
16	to	25	A	A	B	B	B	C	D
26	to	50	A	B	B	C	C	D	E
51	to	90	B	B	C	C	C	E	F
91	to	150	B	B	C	D	D	F	G
151	to	280	B	C	D	E	E	G	H
281	to	500	B	C	D	E	F	H	J
501	to	1200	C	C	E	F	G	J	K
1201	to	3200	C	D	E	G	H	K	L
3201	to	10000	C	D	F	G	J	L	M
10001	to	35000	C	D	F	H	K	M	N
35001	to	150000	D	E	G	J	L	N	P
150001	to	500000	D	E	G	J	M	P	Q
500001	and	over	D	E	H	K	N	Q	R

(Source: Mil-Std 105 Standard Sampling plans)

*Note that General inspection level II is a commonly used level of inspection. Further details on inspection levels are provided in the Mil-Std tables.

75. This lot size is found in the first column of the table. It is included in the "Lot or batch size" 51 to 90. Joe followed across the table to the column headed "General inspection level II" and found the sample size code letter E. Joe then moved to the second table titled Single Sampling Plans for Normal Inspection. (See Table 7-2.) The first column of this table shows the sample size code letters. Moving down to the code letter E and looking in the next column, Joe found the sample size to be 13. Moving straight across the table to the columns headed "Acceptable Quality Levels", Joe found that the acceptance number (Ac) for an acceptable quality level of 1% defective (under the column headed 1.0) is zero when using a sample size of 13. This is the way Joe determined how many balloons to inspect (13) and when a lot should be rejected (when *any* balloons in the sample are defective).

By using acceptance sampling, Joe can be confident that the boxes of balloons he accepts are highly likely to be at the quality level he wants, and the boxes he rejects for sorting are very probably not at the quality level he wants. By inspecting every balloon in the rejected boxes, Joe is maintaining an average quality level of the balloons being shipped, and he knows what this average quality level is.

TABLE 7-2
Single sampling plans for normal inspection.

Sample size code letter	Sample size	0.010 Ac Re	0.015 Ac Re	0.025 Ac Re	0.040 Ac Re	0.065 Ac Re	0.10 Ac Re	0.15 Ac Re	0.25 Ac Re	0.40 Ac Re	0.65 Ac Re	1.0 Ac Re	1.5 Ac Re	2.5 Ac Re	4.0 Ac Re	6.5 Ac Re	10 Ac Re	15 Ac Re	25 Ac Re	40 Ac Re	65 Ac Re	100 Ac Re	150 Ac Re	250 Ac Re	400 Ac Re	450 Ac Re	1000 Ac Re
A	2													0 1			1 2	2 3	3 4	5 6	7 8	10 11	14 15	21 22	30 31		
B	3												0 1			1 2	2 3	3 4	5 6	7 8	10 11	14 15	21 22	30 31	44 45		
C	5											0 1			1 2	2 3	3 4	5 6	7 8	10 11	14 15	21 22	30 31	44 45			
D	8										0 1			1 2	2 3	3 4	5 6	7 8	10 11	14 15	21 22	30 31	44 45				
E	13									0 1			1 2	2 3	3 4	5 6	7 8	10 11	14 15	21 22	30 31	44 45					
F	20								0 1			1 2	2 3	3 4	5 6	7 8	10 11	14 15	21 22								
G	32							0 1			1 2	2 3	3 4	5 6	7 8	10 11	14 15	21 22									
H	50						0 1			1 2	2 3	3 4	5 6	7 8	10 11	14 15	21 22										
J	80					0 1			1 2	2 3	3 4	5 6	7 8	10 11	14 15	21 22											
K	125				0 1			1 2	2 3	3 4	5 6	7 8	10 11	14 15	21 22												
L	200			0 1			1 2	2 3	3 4	5 6	7 8	10 11	14 15	21 22													
M	315		0 1			1 2	2 3	3 4	5 6	7 8	10 11	14 15	21 22														
N	500	0 1			1 2	2 3	3 4	5 6	7 8	10 11	14 15	21 22															
P	800			1 2	2 3	3 4	5 6	7 8	10 11	14 15	21 22																
Q	1250	0 1		1 2	2 3	3 4	5 6	7 8	10 11	14 15	21 22																
R	2000		1 2	2 3	3 4	5 6	7 8	10 11	14 15	21 22																	

(Source: Mil Std 105 Standard Sampling Plans)

SAMPLING RISKS

Before we get into any specifics on how sampling works, we need to point out the risks—yes risks—involved in sampling. What would happen, for instance, if Joe happened to select 13 good balloons from a box of balloons that were mostly defective? Based on the quality of the sample, the box would be shipped! Or what would happen if Joe should pull the only defective balloon from a box? An employee would have to sort through the rest of the box only to find that all the other balloons were good! How can this happen?

Because sampling involves inspecting only a few items in a lot and not every item, a certain amount of risk is involved for both the seller (or producer) of a service, and the customer (or consumer). Both the producer and the consumer bear certain risks that the samples chosen during sampling will not always accurately reflect the quality of the entire lot. Adding to these risks are many factors such as the skill and training of the inspector, the conditions at the workplace, the equipment used when performing the inspection, the number of defects actually in the lots, and many others. These are all part of a subject that involves a mathematical calculation of the chance that defects will not be detected by the inspector. This subject is not covered in this book. We only need to know and understand that errors exist in any plan we devise to estimate, monitor, or control the quality of a service.

Producer's Risk and Consumer's Risk

A producer of a service risks that "good" lots—those that meet the standard for acceptability—will be rejected. This happens when, by chance, the sample from a good lot is defective and the good lot is rejected. This risk is called the *producer's risk*.

On the other hand, there is the risk that "bad" lots—those that don't meet the standard for acceptability—will be accepted. This happens when, by chance, the sample from a bad lot is acceptable and therefore the bad lot is accepted. Although this is a risk the producer of the service takes, it also involves the customer, or consumer of the service, because the customer receives the "defective" service. For this reason it is called the *consumer's risk*.

We'll find out more about how to assess these risks later, but now let's take a look at some areas in which service industries can use sampling.

APPLICATIONS OF SAMPLING PLANS

Sampling plans can be used effectively in many areas of a service industry. They can be used wherever large amounts of the same item or material are handled. Many different materials are used in the production and delivery of a service. Some examples of such materials are printed forms, linen and bedding, food, auto repair parts, etc.

You have learned how to use control charts to tell when something is wrong in the process—when you might be making defects that cause poor quality. But remember: control charts are great tools of quality because they are based on the *sequence* of production. If we no longer know the sequence in which the items have been produced, a control chart loses its effectiveness. When such a condition arises and we want to control the quality level of services or products, we can use a sampling plan.

Sampling plans can be used in:

Health care	For checking incoming shipments of medicines, invoices, patient admittance forms, patient records, and housekeeping records
Insurance	For policy application forms, written policies, and policyholder bills
Banking	For depositor statements and teller records
Equipment service and repair	For repair orders and invoices
Sales of food, clothing, hardware, etc.	For incoming shipments of stock and for sales slips
Restaurants	For incoming food shipments and customer sales slips
Auto service and repair	For service request forms, incoming replacement parts, purchase orders, and customer bills

These are just a few examples of the many areas in which sampling plans can be used in service situations.

SAMPLING PLANS

A sampling plan is simply a set of specific guidelines for sampling and accepting or rejecting material depending on the quality of samples drawn from lots of the material.

Sampling plans are generally attribute plans. That is, when samples are inspected, certain characteristics, or attributes, of the items in the samples are examined to determine whether each item in the sample is "good" or "bad." The attributes you examine, and your standards for acceptability, will vary depending on the kind of service or product you deliver. For instance, if you are in a bank inspecting customer statements, the customer account numbers on the statements could be one of the attributes you inspect in a sample. These account numbers will be either correct or incorrect (if the number is completely missing, it is considered incorrect). Your decision to accept or reject an entire batch of statements will depend on the number of nonconforming items, or defective statements, you find in your sample.

It isn't unusual to check the quality of more than one attribute in a sample. You must, however, know your standards for judging defective attributes and be careful not to change the standards after you've started inspecting. Changing the standards, or adding or removing conditions for acceptability, will change the probability of rejecting or accepting a lot of material and change the risks taken by the producer and the consumer.

When properly used, sampling plans are an alternative to 100% inspection and sorting. In fact, sampling plans may be more effective than 100% inspection. One hundred percent inspection is time consuming and subject to errors because of monotony, mechanical failure, and inspector fatigue. In other words, 100% inspection is not 100% accurate! Besides, it's often not economical to inspect every item in a lot or group. And when inspection involves a destructive test—one that ruins the item being inspected (like opening a can of food)—it *cannot* be performed on every item.

Standard Sampling Plans

Many types of sampling plans have been developed and published over the years, like the one Joe used in the example presented previously. They are generally referred to as Standard Sampling Plans. Many of these plans were developed and published by the United States War Department for use in the production of war material during World War II. Originally known as Joint Army/Navy Standard Sampling Plans, they are known today as the *Military Standard (Mil-Std) 105E* Standard Sampling Tables. These tables are the ones most frequently specified when a standard sampling plan is used. Other tables, however, are also available. (Several of these are referenced in Recommended Readings and Resources.)

You can also design a sampling plan to fit your particular needs. Later in the module, you will see how to do this.

When you look through the various books of sampling plans you will find single sampling plans, double sampling plans, and sequential sampling plans. Double and sequential sampling plans are more complex and cumbersome to use than are single sampling plans. Single sampling plans are by far the most widely used. For this reason we will discuss only single sampling plans in this module.

PARAMETERS OF SAMPLING PLANS

Every sampling plan has parameters that define the plan and determine the quality it tests for. The basic parameters of a sampling plan are the sample size (n), the acceptance number (c), and the lot size (k).

The *sample size (n)* is the number of items you'll inspect from a lot of material in order to decide whether the lot is acceptable.

The *acceptance number (c)* is the maximum number of defective items, or rejects, you'll allow in a sample and still accept the lot without further inspection.

The *lot size (k)* is the total number of items in the group or lot being sampled.

Sampling plans vary in their ability to assess the quality of material. We describe the capability of a sampling plan by using probability numbers. These numbers indicate the chance that a sampling plan will accept material of a given level of quality. They are called *probabilities of acceptance* (P_a).

Probabilities of acceptance are usually expressed as decimals. A probability of 1.0 means 100 in 100, or 100%. A probability of .10 means 1 in 10, or 10%. A probability of .95 in a sampling plan means there is a 95% chance that material of a given quality level will be accepted by that plan.

For example, if you submit lots in which 2% of the items are defective to a sampling plan, the probability of their being accepted depends upon the sampling plan you are using. Some sampling plans are designed to make 2% defective lots acceptable; others will reject such lots. The key to using sampling plans effectively is to select the plan that will best meet your needs.

LIMITATIONS OF SAMPLING PLANS

Even if your sampling plan is designed to accept 2% defective material, the risk always exists that some lots that are less than 2% defective will be rejected and some lots that are more than 2% defective will be accepted. These are the risks associated with using sampling plans, known as the producer's risk and consumer's risk.

The risk of inaccurately assessing the quality of a lot always exists, especially when you are not inspecting every item. We judge the capability of a sampling plan by how well it can detect good and bad lots of material. The better a sampling plan assesses the quality of material, the more it minimizes the risks inherent in sampling.

When judging the effectiveness of a sampling plan, you need to consider the risk you're willing to take for bad acceptance decisions. You also need to consider the cost of using the sampling plan and how this cost compares with the cost of 100% inspection. In this module you'll learn how to look at some of these costs. But you should also consider that there are other costs involved with sampling, such as the cost to correct a defect, the loss of a customer, etc.

SAMPLING TECHNIQUES

Whatever sampling plan you use, the accuracy of your sample can be affected by the technique you use to obtain the sample. A faulty sampling technique can completely defeat a good sampling plan.

Representative Sampling

When using a sampling plan, you need to select samples that have the best chance of reflecting the quality of the entire lot from which they are drawn. Samples like this are called *representative samples*. In order to obtain representative samples, you must use sampling techniques that minimize the occurrence of *sampling biases*. Possible biases in sample selection include:

1. Sampling from the same location in all lot containers.
2. Not drawing samples from areas that are hard to reach.
3. Preinspecting items before sample selection and taking only those that look defective (or only those that *don't* look defective).
4. Using a "pattern" of sampling without knowing if this is necessary.

Establishing sample selection procedures and following them at all times will minimize the occurrence of sampling biases. This helps to ensure proper sampling so that the probabilities predicted by your sampling plan are the ones you actually get.

Random Sampling

In a *random sample*, every item in the lot has the same chance of being selected as a part of the sample. A random sample is representative; a sample that has been influenced by biases may not be.

When you want random samples, you need to know if the process or the handling system used to produce and deliver the lot you'll be sampling has been disrupted from its sequence of production. Many processes and handling systems are systematic—that is, they maintain a lot's sequence of production. Other processes and handling systems, however, are very good "mixers"—they disrupt the order in which a lot was produced.

We usually group items for sampling inspection in what is called an *inspection lot*. The items are generally in some form of container or containers. They may just be in piles of paper forms. If the containers are dumped helter-skelter and thoroughly mixed, and the items are passed by the inspector one at a time, any item is as likely as any other to be the next one to pass the inspector. These items are relatively randomized within the inspection lot.

If your inspection lot already is randomized, it's easier for you to select a representative sample. If you cannot tell whether the material you'll be sampling has been randomized, you need to take special measures to select a random sample. You can do this by using random numbers to choose your samples. Tables of random numbers are readily available in various books on statistics, and many calculators can generate random numbers.

Using Random Numbers

In order to use random numbers to select your sample, you must be able to identify each item in your inspection lot by a number. If, for example, your inspection lot is found in file drawers, you can use three-digit random numbers. (But be sure your file drawers are all about the same size.) The first digit can identify the file drawer; the second digit, the file folder; the third, a specific item in the file folder. Using this procedure will help you pick a random, or representative, sample if your inspection lot was formed or packed in a systematic or nonrandom manner.

Following this procedure, however, requires descipline. You must closely monitor the sampling, or sample selection can easily become biased.

Now that you know the basics of acceptance sampling, you are ready to learn how to look at a sampling plan, evaluate its effectiveness and decide whether it's a good plan for your situation. Remember, a sampling plan is just a set of guidelines for sampling that includes the sample size, acceptance number, and lot size.

OPERATING CHARACTERISTIC CURVES

We can visually portray the risks associated with any sampling plan by using curves called the operating characteristic (OC) curve and the average outgoing quality (AOQ) curve.

The OC Curve

An *operating characteristic curve* is a curve that shows, for a given sampling plan and a given lot percent defective, the probability that such a lot will be accepted by the sampling plan.

Each sampling plan has its own *OC curve*. Drawing a sampling plan's OC curve can help you "see" the risks involved with using that plan. We'll do that, but first let's look at some OC curves to see how they work and what they can tell us.

Remember, the two kinds of risks in acceptance sampling are:

1. Producer's risk—the risk or probability of *rejecting* a *good* lot. The producer's risk is often denoted by the symbol α (read as "alpha").
2. Consumer's risk—the risk or probability of *accepting* a *bad* lot. The symbol β (read as "beta") is often used to denote the consumer's risk.

If we could design an ideal sampling plan, both of these risks would be eliminated. For example, if management has decided that 1% defective lots are acceptable, ideally we will always accept lots that are 1% defective or less. Likewise, we will always reject lots that are more than 1% defective. The OC curve for this ideal sampling plan is shown in Figure 7-1.

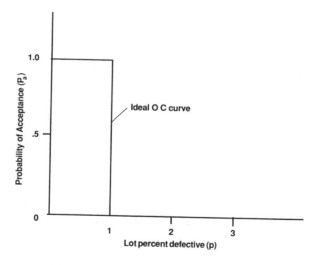

Figure 7-1. Ideal operating characteristic curve.

Notice that in this ideal case, the producer of the material is at zero risk that lots less than 1% defective will be rejected, because the probability of accepting good lots is 1.0, or 100%. The producer and consumer of the material are at zero risk that material greater than 1% defective will be accepted, because the probability of accepting such material is zero.

If we could do 100% inspection perfectly, the curve shown in Figure 7-1 would be attainable. Using a sampling plan makes probabilities like those shown in Figure 7-1 unattainable. But by selecting the proper sampling plan, we can strike a balance between the risks of sampling and the drawbacks of 100% inspection. Figure 7-2 shows a comparison between an ideal OC curve and an attainable or practical OC curve.

In Figure 7-2, an example of an "actual" OC curve is drawn using a broken line. The ideal OC curve is shown using a solid line. If 1% defective material is acceptable, we can find the producer's and consumer's

Figure 7-2. Comparison of ideal operating characteristic curve with attainable curve.

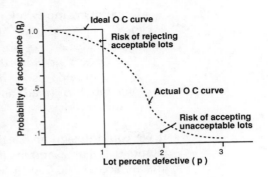

risks for both these sampling plans by looking at the OC curves. Let's see how.

HOW TO INTERPRET AN OC CURVE

In Figure 7-2, the horizontal axis represents the quality level of the material you'll be sampling. In this example, 1% defective is considered acceptable. To find the producer's risk of the "actual" plan, we start at the 1% mark on the horizontal axis and move straight up the line to the point where we hit the actual OC curve. From this point, we move straight across to the left and intersect the vertical axis labeled "Probability of acceptance (P_a)." The value represented by this point of intersection is about .85. This means the probability of this plan accepting a 1% defective lot is .85 or 85%.

Producer's Risk

If lots coming are 1% defective, 85% of them will be accepted by the plan. It follows that the remaining 15% will be rejected even though they are also 1% defective. This 15% is the producer's risk for the plan. It is the risk of acceptable lots being rejected. The producer's risk for this actual plan, then, is about .15 or 15%.

Consumer's Risk

The consumer's risk is defined as the probability of accepting a bad lot of material. Let's say a bad lot of material is one that is 2.25% defective. We can find the 2.25% defective point on the horizontal axis in Figure 7-2 and move straight up to the point where we meet the actual OC curve. From this point, we move straight across to the left and intersect the vertical axis labeled "Probability of acceptance (P_a)." The value represented by this point of intersection is about .10. This means the probability of this plan accepting 2.25% defective lots is .10 or 10%. This is the consumer's risk for this plan. It is the risk of bad or unacceptable lots being accepted when using the plan represented by the actual OC curve.

For lots of a stated quality, an OC curve will show us the probability that these lots will be accepted by a given sampling plan. Stated another way, the OC curve shows the long-run percentage of lots that would be accepted if a great many lots of any stated quality were submitted for inspection using the given sampling plan.

By looking at the "actual" OC curve in Figure 7-2, we can see that as the quality of material worsens (the percent defective increases), the probability of accepting this material decreases.

Capability

A sampling plan whose OC curve closely resembles the ideal curve has a greater capability for discriminating between lots of acceptable and unacceptable quality than a plan whose OC curve is far from the ideal. We can demonstrate this by comparing two OC curves with an ideal curve (Figure 7-3).

In Figure 7-3, the OC curve for sampling plan A looks more like the ideal curve than the OC curve for sampling plan B. The probability of accepting 1% defective material is the same for both plan A and plan B. For material that is 2% defective, however, plan A shows a probability of acceptance that is very small and plan B shows a probability of acceptance that is very high. This means that sampling plan A will do a better job than sampling plan B when it is used to sample lots of varying quality levels.

DEVELOPING AN OC CURVE

You can better understand the use of an OC curve by actually developing the OC curve for a sampling plan. The procedure we will use to develop an OC curve is one that uses very little mathematics but does

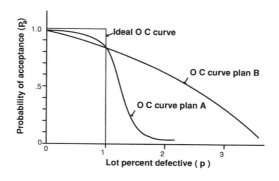

Figure 7-3. Comparison of two OC curves with ideal curve.

result in a curve that is a good approximation of the true curve for the sampling plan it represents.

Since each sampling plan has a different OC curve, we must first select a plan. For our example we have selected a plan with a sample size (n) of 15 and an acceptance number (c) of zero. The lot size (k) from which the sample is drawn must be large—100 or more. Be sure your lots are at least 100 in size. With this information we can now develop a table for drawing the OC curve.

The four-column table we have developed is shown in Table 7-3. The first two columns are headed "LOT DEFECTIVE," the third column is headed "np," and the fourth column is headed "PROBABILITY OF ACCEPTANCE (P_a)."

The procedure for developing the OC curve is as follows:

Step 1. List lot defective values in table form.

In the first column we have listed the lot percent defective values as a percentage (%), since this is the usual way we talk about the quality of material. The second column, headed "p (decimal form)" lists the same values as decimals. To obtain these values, you just divide the percent defective value (%) by 100. The lot defective must be in decimal form for us to calculate a number we need to obtain the probability of acceptance for each lot defective value. When you become familiar with this procedure, the decimal form is all you will need to list in your table.

Values from 0 to 20 are listed. The values chosen will determine points on the OC curve. If we should find that we cannot draw a smooth curve with the number of values selected, we may have to add more values to columns one and two.

Step 2. Determine the values for np.

The third column of the table is headed "np." To calculate the values in this column, multiply the lot defective (p) by the sample size (n). The sampling plan in our example has a sample size (n) of 15. The first value in this column is zero, since the first number in the second column is zero ($0 \times 15 = 0$). The second value in this column is .15 ($.01 \times 15 = .15$); the third value is .30 ($.02 \times 15 = .30$); and so on down to the last value, which is 3.00 ($.20 \times 15 = 3.00$).

Step 3. Determine the probability of acceptance (P_a).

The fourth column lists the probability of acceptance (P_a) for each lot percent defective. We obtained these values by using some complicated mathematics. Fortunately, you do not have to perform these calculations. You can find them in Table 7-4, as well as in many standard statistical

TABLE 7-3
Table for developing an OC curve.

LOT DEFECTIVE			PROBABILITY OF ACCEPTANCE
%	p (decimal form)	np	(P_a)
0	0	0	1.00
1	.01	.15	.86
2	.02	.30	.74
3	.03	.45	.64
4	.04	.60	.55
5	.05	.75	.47
6	.06	.90	.41
7	.07	1.05	.35
8	.08	1.20	.30
9	.09	1.35	.26
10	.10	1.50	.22
11	.11	1.65	.19
12	.12	1.80	.17
13	.13	1.95	.14
14	14	2.10	.12
15	.15	2.25	.11
20	.20	3.00	.05

SAMPLING PLAN

n = 15   n = Sample size

c = 0   p = Lot percent defective in decimal form

k = Large

reference books. The probability of acceptance may also be obtained from cumulative probability curves found in many statistical reference books.

We will use Table 7-4 to find the probability of acceptance values for our example. The first column is headed "np" and the numbers under

TABLE 7-4
Probability of acceptance (P_a) values.
(Probability of c or less defects in a sample that has an average number of defects equal to np; based on Poisson's Exponential Binomial Limit.)

np \ c	0	1	2	3	4	5	6	7	8	9
.02	.98	1.00								
.04	.96	.99	1.00							
.06	.94	.99	1.00							
.08	.92	.99	1.00							
.10	.91	.99	1.00							
.15	.86	.99	.99	1.00						
.20	.82	.98	.99	1.00						
.25	.78	.97	.99	1.00						
.30	.74	.96	.99	1.00						
.35	.71	.95	.99	1.00						
.40	.67	.94	.99	.99	1.00					
.45	.64	.93	.99	.99	1.00					
.50	.61	.91	.99	.99	1.00					
.55	.58	.89	.98	.99	1.00					
.60	.55	.88	.98	.99	1.00					
.65	.52	.86	.97	.99	.99	1.00				
.70	.50	.84	.97	.99	.99	1.00				
.75	.47	.83	.96	.99	.99	1.00				
.80	.45	.85	.95	.99	.99	1.00				
.85	.43	.79	.95	.99	.99	1.00				
.90	.41	.77	.94	.99	.99	1.00				
.95	.39	.75	.93	.98	.99	1.00				
1.00	.37	.74	.92	.98	.99	.99	1.00			
1.10	.33	.70	.90	.97	.99	.99	1.00			
1.20	.30	.66	.90	.97	.99	.99	1.00			
1.30	.27	.63	.86	.96	.99	.99	1.00			
1.40	.25	.59	.83	.94	.99	.99	.99	1.00		
1.50	.22	.56	.81	.93	.98	.99	.99	1.00		
1.60	.20	.53	.78	.92	.98	.99	.99	1.00		
1.70	.18	.49	.76	.81	.97	.99	.99	1.00		
1.80	.17	.46	.73	.89	.96	.99	.99	.99	1.00	
1.90	.15	.43	.70	.88	.96	.99	.99	.99	1.00	
2.00	.14	.41	.68	.86	.95	.98	.99	.99	1.00	
2.20	.11	.36	.62	.82	.93	.98	.99	.99	1.00	
2.40	.09	.31	.57	.78	.90	.96	.99	.99	.99	1.00
2.60	.07	.27	.52	.74	.88	.95	.98	.99	.99	1.00
2.80	.06	.23	.47	.69	.85	.94	.98	.99	.99	1.00
3.00	.05	.20	.42	.65	.82	.92	.97	.98	.99	1.00
3.20	.04	.17	.38	.60	.78	.90	.96	.98	.99	1.00
3.40	.03	.15	.34	.56	.74	.87	.94	.98	.99	1.00
3.60	.027	.126	.303	.515	.706	.844	.927	.969	.988	.996
3.80	.022	.107	.269	.473	.668	.816	.909	.960	.984	.994
4.00	.018	.092	.238	.433	.629	.785	.889	.949	.979	.992
4.20	.015	.078	.210	.395	.590	.753	.867	.936	.972	.989
4.40	.012	.066	.185	.359	.551	.720	.844	.921	.964	.985
4.60	.010	.056	.163	.326	.513	.686	.818	.905	.955	.980

the heading are np values from .02 to 4.60. The row across the top of the table lists the c values, or acceptance numbers, of sampling plans. (Note that if the percent defective of the lots inspected were zero, the lots would always be accepted and the probability of acceptance would be 1.0. See the top line of the fourth column in Table 7-3.)

To find the probability of acceptance values for the np values, we enter the first column of the table at the np value and move across to the column headed by the acceptance number of our sampling plan. In our example, the acceptance number is zero. The first np value in Table 7-3 is .15. So we enter Table 7-4 at .15 and move across to the c=0 column. We find that the probability of acceptance value is .86. (This is the entry in Table 7-3 for the second line in the fourth column.) The next np value in Table 7-3 is .30 and in Table 7-4 we find the P_a is .74. We proceed in this way down the np column in Table 7-3 until we come to the np value of 1.05 and find it is not listed in Table 7-4. When this happens, we look at the values for 1.00 and 1.10. Since 1.05 is halfway between those two values, we can say that the P_a for 1.05 is halfway between the P_a values for 1.00 and 1.10. This is .35. Follow the same procedure for np values of 1.35, 1.65, 1.95, and 2.25. You will find 1.95 will have a P_a of .145, but since Table 7-4 shows numbers to only two decimal places, the P_a value is shown as .14. The np value 2.25 would also have three decimal places, but it has been shortened to .11. The precision of these values is adequate for our development of the OC curve.

Step 4. Draw the OC curve.

To draw the OC curve, we plot the lot percent defective values in the first column of Table 7-3 against the corresponding probability of acceptance values in the fourth column and connect the plotted points with a line. Figure 7-4 shows this.

The first lot percent defective value in Table 7-3 is zero and the cor-

Figure 7-4. Drawing the OC curve.

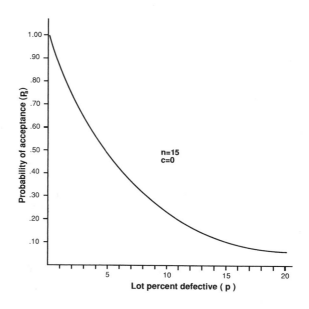

responding probability of acceptance value is 1.0, so we start by finding the zero percent defective value on the horizontal line of the chart in Figure 7-4. From here we move up the vertical line, which is the probability of acceptance (P_a), to the P_a value of 1.0 and place a small x or dot. The next lot percent defective value is .01, or 1% and the P_a value is .86. We move along the horizontal line to the first hatch mark, which represents 1% defective, and move up along the vertical line to a point directly above the 1% defective mark and straight across from .86 on the probability of acceptance line. (This is just a little beyond halfway between .80 and .90.) We mark this point with a small x or dot. When we have marked all the points corresponding to the percent defective and P_a values, we can draw a smooth curve connecting them. This is our OC curve for the sampling plan $n = 15$, $c = 0$.

ACCEPTABLE QUALITY LEVEL AND LOT TOLERANCE PERCENT DEFECTIVE

Two points on every OC curve have special significance: the points that represent a sampling plan's *acceptable quality level (AQL)* and *lot tolerance percent defective (LTPD)*. These points relate directly to the producer's and consumer's risks. This is shown in Figure 7-5.

We speak of the quality of material in terms of percent defective, whether the material is hamburgers, order forms, or incoming shipments of supplies. The AQL is the quality, expressed as percent defective, of material lots that is considered good or acceptable. When lots of this quality are submitted to a sampling plan we want them to be accepted. However, we already know that because of variation in sampling, the risk exists that some of these lots will be accepted and some will be rejected, even though they have the same quality. This is the producer's risk; it is customarily set at 5%. Thus, the acceptable quality level of a sampling plan is that lot quality which will be accepted 95% of the time; we as a producer run a risk of 5% that lots of that quality will be rejected. Usually, we first decide on an AQL and then design or select a sampling plan that will accept the AQL 95% of the time.

The LTPD is the quality of material lots we would rather not accept. But, again, we know the risk exists that this material will be accepted. This is called the consumer's risk, even though we as producers also do not want to accept material of this quality. It should be very low. The probability of acceptance of the LTPD is customarily set at 10%. We can think of the LTPD as the limit of acceptable quality. It is the level of quality a particular sampling plan will reject 90% of the time.

The sampling plan shown by the OC curve in Figure 7-5 has an AQL of 1.5% defective and an LTPD of 6% defective. When a lot with 1.5% defective material is sampled using this plan, its probability of acceptance is about 95%. This can be seen by following the broken line on the chart

Figure 7-5. OC curve showing producer's risk and consumer's risk.

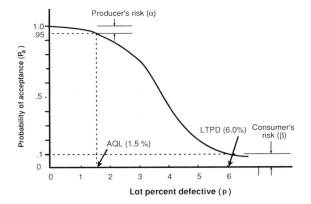

starting at the 1.5 point on the horizontal, or lot percent defective line, moving straight up until we come to the OC curve, and moving from there across to the probability of acceptance line. We see that the value on that line is .95. Therefore, using this plan, the producer runs a 5% risk that each of his acceptable lots of 1.5% defective material will be rejected.

When a lot with 6% defective material is sampled using this plan, its probability of acceptance is about 10%. Using this plan we run a 10% risk that we will accept lots that are 6% defective. This can be seen by entering the chart at the 6% defective value on the horizontal line, moving up to the OC curve, and from there moving across to the probability of acceptance line where we see that the probability of acceptance is .10 or 10%.

When evaluating the effectiveness of a sampling plan, we must decide whether the sampling risks incurred by the plan are acceptable. In other words, do we want a sampling plan that will accept 1.5% defective material 95% of the time and reject it 5% of the time? A plan that will accept 6% defective material 10% of the time and reject it 90% of the time?

The AQL and LTPD are important points to consider when evaluating a sampling plan. If the AQL or LTPD of our sampling plan is not right—the percent defective material being accepted is too high or material of questionable quality is being accepted more often than we want—we can alter our sampling plan by changing its sample size or acceptance number. Changing any of the parameters of a sampling plan will also change the risks associated with that plan.

CHANGING THE ACCEPTANCE NUMBER

An acceptance number (c) of zero means we must reject the lot if we find any defectives in our sample. An acceptance number (c) of 2 means we will accept the lot if we find zero, 1, or 2 defective items in our sample. Only when we find more than 2 defectives in the sample will we reject the lot.

Increasing a sampling plan's acceptance number will increase a sample's probability of acceptance. If you think about it for a minute, this makes perfect sense. The more rejects you allow in an acceptable sample, the higher the percentage of lots you'll accept.

The OC curves in Figure 7-6 show that the probability of accepting lots of a particular quality level increases as the acceptance number (c) increases. In this case, the sample size (n) of the plan is 20. This principle will still apply, however, if the sample size is changed.

Figure 7-6 shows the consequence of an inspector inadvertently increasing or decreasing the acceptance number for a specified sampling plan. For example, an 8% defective lot has about a 50% chance of being accepted when the acceptance number is 1. The inspector erroneously decides that something about an item in a sample is not a defect when it actually is. He or she has increased the acceptance number from 1 to 2 because if that condition appears in a sample that contains another defect, the lot will be accepted even though the sample actually contained 2 defects. Increasing the acceptance number to 2 would give that lot about an 80% chance of being accepted!

Likewise, when the inspector decides something about an item is a defect when it actually is not, it decreases the probability of acceptance of a lot. If such a condition exists in a sample, then the sampling plan does not allow any more defects in order to accept the lot. This effectively makes the acceptance number zero instead of 1. Referring to Figure 7-6 we can see that lots containing 8% defective material would have a 20% chance of being accepted.

When using sampling plans, it is very important for all inspectors to know exactly what constitutes a defect and what does not. If one inspector views an item as defective and includes it with the total defects in the sample and another inspector does not, the probability of acceptance will change depending on who is inspecting the material rather than on the quality of the material being inspected. Inspector instructions must be clear and concise.

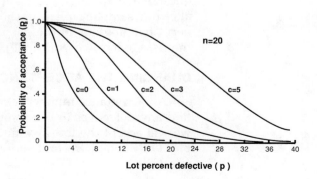

Figure 7-6. Effect on OC curve of changing the acceptance number, c.

CHANGING THE SAMPLE SIZE

Increasing a sampling plan's sample size will reduce the plan's probability of acceptance. Again, if you think about it for a minute, this makes sense. The more items we include in our sample, the more likely we are to find any defective items in the lot.

The OC curves in Figure 7-7 show this. For example, the probability of accepting 12% defective lots decreases as the sample size increases. The probability of acceptance using sample sizes of 8, 13, and 20 are approximately .75, .55, and .30, respectively.

From looking at these OC curves, we can see that sampling plans using larger sample sizes do a better job accepting good quality and rejecting bad quality than do plans with smaller sample sizes. These curves demonstrate the need to "stick to the plan" and not adjust the specified sample size in the interest of economy or time.

Figure 7-7. Effect on OC curve of changing the sample size, n.

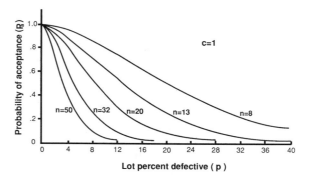

Keeping the Sample Size at a Fixed Percentage of a Lot

Some companies make the error of using a sample that is a fixed proportion of the incoming lot. For example, a company may decide that it will inspect 10% of every incoming lot. If they find no rejects in the sample they will accept the lot.

Using this inspection policy indicates that managers do not understand how changing the sample size affects the probability of acceptance.

In the plans shown by the OC curves in Figure 7-8, the sample size is 10% of the lot size.

The top OC curve shows the probability of acceptance if incoming lots contain 50 items (k=50). The lot size (k) is 50, the sample size (n) is 5 (10% of 50), and the acceptance number (c) is zero. The curve shows that 2% defective lots will be accepted approximately 91% of the time.

On the other hand, the bottom OC curve shows what happens if incoming lots contain 1,000 items and the sample size is maintained at 10% of the lot size. The sample size (n) in this case will be 100. Now,

Figure 7-8. Effect on OC curve of keeping the sample size a fixed percentage (10%) of lot size.

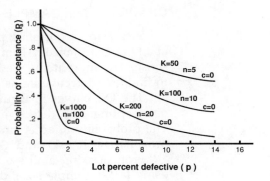

only 13% of the 2% defective lots will be accepted! Likewise, the OC curves for lot sizes (k) of 100 and 200 items show that the probabilities of accepting these 2% defective lots are approximately .82 and .67 respectively.

What's happening here? The probability of accepting lots of the same quality is changing with the *size* of the lots and not with the *quality* of the lots! The protection against rejecting good material and accepting bad material is changing with each change in lot size. Neither the producer nor the consumer is well served by such practice. This is why we oppose the practice of taking a fixed percentage of the lot as the sample.

AVERAGE OUTGOING QUALITY

Sampling plans are not usually seen as tools for process control. However, a well-designed sampling plan can be an asset to any company using SPC and having an active program of continuous improvement. Such a company uses SPC to control or maintain the quality levels determined by the capabilities of the current processes. Continuous improvement activities work to improve the capabilities of the processes and thereby improve quality.

Many companies recognize that their current processes do not have the capabilities they must have to meet the wants, needs, and expectations of the customer. Until the required capabilities are achieved through continuous improvement activities, the use of sampling plans can help maintain the average outgoing quality of services at a level acceptable to the customer.

Average outgoing quality (AOQ) is the average quality of material after it has been submitted to a sampling plan and defective items in rejected lots have been sorted out and corrected.

If the quality of the lots being sampled and inspected is very good, most of the lots will be accepted and the small number of lots rejected will be corrected. The AOQ will be very good. If the quality of the lots being sampled and inspected is very bad, most of the lots will be rejected and corrected. The defects found in the rejected lots will be fixed or replaced, and the corrected lots will be merged with the lots that were accepted. The quality level of these corrected lots is then averaged with the quality level of the accepted lots. Therefore, the AOQ of the combined lots will be very good. As it turns out, if the incoming lot quality is intermediate, or in between very good and very bad, the AOQ will be poorer, or have a higher average percent defective, than in the previous two cases.

We will show how this works with an example.

DEVELOPING AN AOQ CURVE

To show how the average outgoing quality (AOQ) varies with the quality of the lots being sampled, we can plot a curve called the AOQ curve. We can develop the AOQ curve for the sampling plan in Figure 7-9 using the plan's OC curve. To illustrate this, suppose we inspect 100 lots using this sampling plan. Each lot contains 500 items, or pieces, and all of the lots are 1% defective.

Figure 7-9. An operating characteristic (OC) curve.

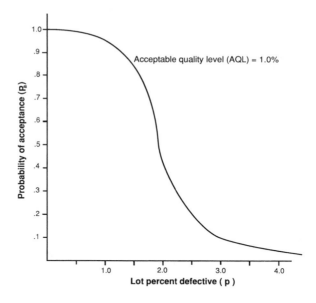

First, we will determine the average outgoing quality of the 100 lots by following the steps shown in Figure 7-10.

Block A represents the 100 lots being submitted to the sampling plan. Each lot contains 500 items, so a total of 50,000 items is involved. (Remember that not all of these items are inspected, however, just those selected for the samples.) Since all the lots are 1% defective, 500 of the 50,000 items are defective. The OC curve in Figure 7-9 shows us that according to this plan, 1% defective material has a .95 probability of

Figure 7-10. Determining the average outgoing quality for 1% defective material.

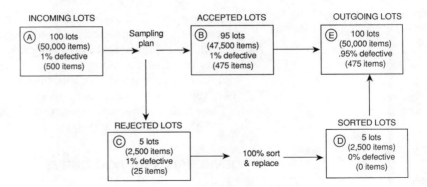

acceptance. This means we will accept 95 of the 100 lots and reject 5 of them.

Block B represents the 95 lots we've accepted. These 95 lots are made up of 47,500 items. One percent of these items, or 475, are defective.

Block C represents the 5 lots we've rejected. These 5 lots are made up of 2,500 items. One percent of these items, or 25, are defective. These 5 lots are screened and the 25 defective items are replaced or corrected.

Block D represents the 5 lots that have been corrected. These lots are now zero percent defective: they contain no defective items. When these 5 screened lots are combined with the 95 lots we accepted, as shown in Block E, you can see that we now have 100 lots (or 50,000 items) which contain a total of 475 defective items. The total percent defective for the 50,000 items is now 475/50,000 or .95%. This is the average outgoing quality of these lots. This total represents one point on our AOQ curve.

Before going on with this example, we should note that we are not taking into account any effect inspection errors would have on the average outgoing quality of the lots.

The development of a second point on our AOQ curve is shown in Figure 7-11. To find this point, suppose we are using the same sampling plan to inspect 100 lots that are 3% defective. Block F represents the 100 incoming lots (50,000 items in all). Since these lots are 3% defective, they contain 1,500 defective items. Our OC curve in Figure 7-9 shows us that 3% defective material has a probability of acceptance of .10 (or 10%). This means we will accept 10 of the 100 lots and reject 90 of them.

Block G represents the 10 lots we've accepted. These 10 lots contain 5,000 items—150 of them defective. The 90 lots we've rejected are represented by Block H. These lots are made up of 45,000 items, and 3% of them, or 1,350, are defective. When the 90 lots are sorted and the defective items corrected or replaced, we have 45,000 items that are zero percent defective (Block I). The sorted lots are combined with the accepted lots as shown in Block J, giving us 100 outgoing lots of 50,000 items—150 of them defective. The average outgoing quality, or total percent defec-

Figure 7-11. Determining the average outgoing quality for 3% defective material.

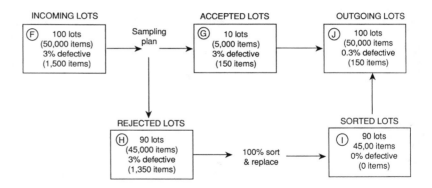

tive, of these lots is now 150/50,000 or .3%. Even though a very large number of the lots were rejected and required sorting and replacement or correction, the average outgoing quality of these lots is very good as a result. This is the second point on our AOQ curve. These points are plotted in Figure 7-12.

This example demonstrates the concept very well, but it is not necessary to follow the above procedure to fully develop the AOQ curve.

Figure 7-12. Average outgoing quality (AOQ) curve.

Column 1 Lot Fraction Defective	Column 2 Probability of Acceptance (P_a)	Column 3 Average Outgoing Quality (AOQ) AOQ = Column 1 times Column 2
.5	.98	.49
1.0	.95	.95
1.5	.78	1.17
2.0	.50	1.00
2.5	.24	.60
3.0	.10	.30
3.5	.06	.21

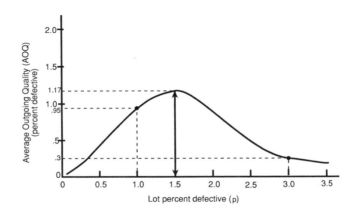

The various points on the curve can be calculated as shown in the table in Figure 7-12. The lot percent defective values are shown in column 1 of the table. The probability of acceptance for each of the values in column 1 is obtained from the OC curve for the sampling plan (Figure 7-9) and recorded in column 2. The average outgoing quality (AOQ) for each percent defective is shown in column 3. These values are obtained by multiplying the percent defective value in column 1 by the probability of acceptance value in column 2.

Using the values in columns 1 and 3, you can draw the AOQ curve, as shown in Figure 7-12. At each lot percent defective value on the horizontal line of the chart, we move up to a point straight across from the corresponding AOQ which we obtain from column 3 and place a small x or dot. When all the points are placed on the chart we connect them with a smooth curve—the AOQ curve.

Locate the high point on the AOQ curve. In this case, it is 1.17% defective (1.2% when rounded off) and it occurs when 1.5% defective material is submitted to the sampling plan. Any material with a percent defective smaller or greater than 1.5% will have an average outgoing quality less than 1.2% defective. This is called the *average outgoing quality limit (AOQL)*. The AOQL represents the poorest long-term average quality level that will be obtained, regardless of the quality level of incoming lots of material, when using a sampling plan with the OC curve shown in Figure 7-9.

The AOQL is the long-term average quality limit, provided that rejected lots are sorted and defective items corrected or replaced with good items.

Using sampling plans in the manner just described, we can know with confidence the poorest long-term quality that might be attained when any items leave our control at any stage in the process of producing a service for the customer. A sampling plan is not a statistical quality control tool that identifies assignable causes; but it is a tool we can use for protection during those times when processes are not fully capable of producing defect-free items and during those times when our control systems break down and assignable causes are present.

SELECTING A SAMPLING PLAN

Sampling plans are often used in situations where you have no information about the conditions present during the production of the items sampled. You may not know whether they were produced in a state of statistical control. In such cases, you should consider the cost or consequences if defective items are accepted into the process. If you expect the consequences to be severe, consider a plan with a low LTPD.

Remember, the AQL (acceptable quality level) is related to the producer's risk, and the LTPD (lot tolerance percent defective) is related to the consumer's risk. The AQL is the quality level that will be accepted 95% of the time and rejected 5% of the time. The LTPD is the quality level that will be accepted by the sampling plan 10% of the time, even though it is not wanted by the consumer (or the producer). The terms producer's risk and consumer's risk are not intended to denote an "us" and "them" relationship but merely to indicate whether the emphasis of the sampling plan lies with the AQL or the LTPD.

Now, you must decide whether it's more important for your material to be the quality of the AQL (the producer's risk) or the LTPD (the consumer's risk). When making this decision, consider the consequences of defective lots as high in percent defective as the LTPD being accepted, or the consequences of acceptable lots as low in percent defective as the AQL being rejected. Don't forget to consider the costs involved in inspecting the lots! (More on this later.)

Use an OC curve when selecting a sampling plan because it will help you "see" the risks associated with the plans you are considering. As we have seen, an ideal sampling plan is one with an OC curve that is a vertical line at the LTPD. With such a plan, the AQL would be the same as the LTPD, and therefore the producer's and consumer's risks would be reduced to zero. We already know this is an ideal that cannot be achieved in practice. But we can find a plan that minimizes our risks as much as possible.

For example, suppose management of your company decides that 2.2% is the maximum percent defective (LTPD) your company can accept in materials you purchase from outside vendors. The ideal sampling plan would reject all lots greater than 2.2% defective and accept all lots less than 2.2% defective. But we know this goal is not possible using any sampling plan.

It is possible, however, to find a sampling plan that will accept 2.2% defective material only 10% of the time (the consumer's risk). One such plan is $n=100$, $c=0$. This plan will give you the protection you want against acceptance of 2.2% defective lots. (See Figure 7-13.)

This plan's OC curve is very restrictive for the producer of the material. Inspection lots that are only 1% defective will be accepted just 37% of the time, and lots that are 0.5% defective will be accepted only 61% of the time. In fact, the AQL for this plan—the percent defective that will be accepted 95% of the time—is about .1% defective!

A plan like this will probably reject large amounts of acceptable material. It is usually not in the best interest of the consumer or the producer to use a sampling plan that rejects excessive amounts of acceptable material in order to protect against acceptance of unwanted material.

As shown earlier, decreasing the sample size will make an OC curve

Figure 7-13. OC curve for a sampling plan: n = 100, c = 0.

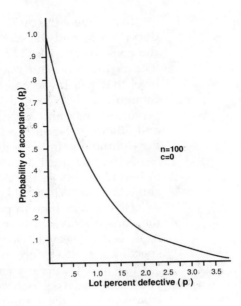

less steep. (See Figure 7-7.) Also, increasing the acceptance number of a sampling plan will increase the AQL. (See Figure 7-6.)

A sampling plan that may be more satisfactory for the producer is n = 240, c = 2. In Figure 7-14, the OC curve for this plan is compared with the OC curve for the n = 100, c = 0 plan. As you can see, the LTPD for these two plans is the same but the AQLs are quite different.

When you consider the sample size, consider the cost of operating the sampling plan. Keep in mind that sample inspection is not the only cost involved in using a sampling plan; the cost of screening or sorting rejected lots must be considered. Weigh these costs against the amount of protection the plan gives the consumer and the producer. Sometimes you will have to make compromises to balance the protection given by the sampling plan with the costs of operating the plan. In this book we only look at the sampling costs; we do not look at the cost of correcting a document, of customer ill will, etc.

Many of these are economic decisions that must be made by the management of your company. Management must consider many factors, including the competition you face in the marketplace, the current level of capability of your processes, the expectations for improvement in quality, and many others. It is not the objective of this book to address those factors. This book provides statistical tools and techniques useful in the daily operation of your company and it provides information management can use to help in making those decisions.

As shown previously, if all rejected lots are inspected 100% and the

Figure 7-14. OC curves for two sampling plans: n = 100, c = 0; and n = 240, c = 2.

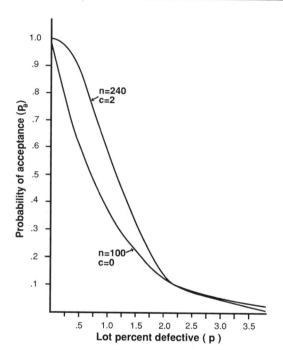

defective items replaced or corrected, we can determine the average outgoing quality. In addition, we can determine the average number of items inspected per lot if we know the average incoming quality or can estimate it. The average number of items inspected per lot of incoming material can be related directly to the costs of using a particular sampling plan. Let's work through an example of how to do this by comparing two sampling plans. We will use the sampling plans represented by the OC curves in Figure 7-14.

Figure 7-15 compares the average number of items inspected for two different sampling plans. For purposes of calculation, suppose that both plans are sampling 100 lots of 1,000 documents that are 1% defective. The first sampling plan calls for a sample size (n) of 100 and an acceptance number (c) of zero. When 100 lots are submitted to this plan, we can expect 37 of them to be accepted. (See the OC curve for n = 100, c = 0 in

Figure 7-15. Comparison of two sampling plans.

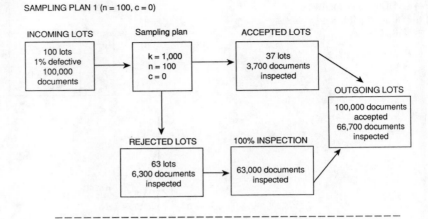

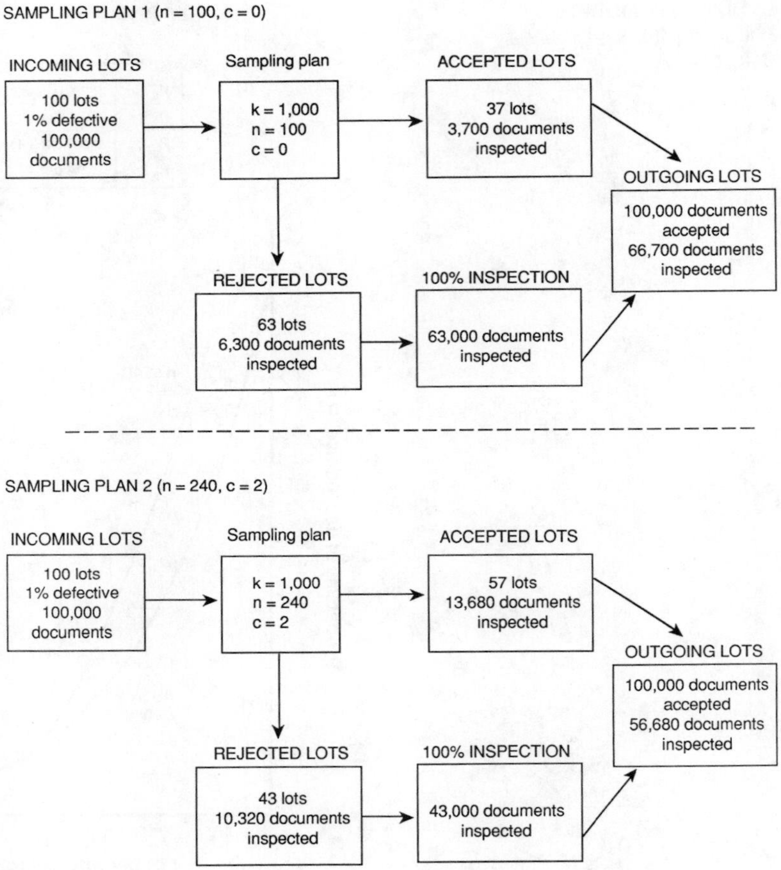

Figure 7-14). In accepting these 37 lots, we will have to inspect 3,700 documents.

The remaining 63 lots will be rejected, inspected 100%, and either corrected or replaced. This means that in addition to the 3,700 documents inspected in the accepted lots, we will also have to inspect another 63 lots, or 63,000 documents. If we use this sampling plan, we will have to inspect a total of 66,700 of the 100,000 documents submitted to the plan.

The other sampling plan shown in Figure 7-15 calls for a sample size (n) of 240 and an acceptance number (c) of 2. When 100 lots are submitted to this plan, we can expect 57 lots to be accepted. (See the OC curve for $n=240$, $c=2$ in Figure 7-14.) In accepting these 100 lots, we will have to inspect 13,680 documents. The remaining 43 lots, or 43,000 documents, will be rejected, inspected, and either corrected or replaced. The total number of documents we will have to inspect using this plan is 56,680.

If we compare the total number of documents we will have to inspect using these sampling plans, we can see that using a smaller sample size is not necessarily a more economical approach. The sampling plan using a 100-piece sample resulted in the inspection of 66,700 documents whereas the plan using a 240-piece sample resulted in the inspection of 56,680 documents—a difference of 10,020 documents!

When using the above approach to select a sampling plan, you should know the average incoming quality of the material to be inspected. You can select and use a sampling plan without knowing the quality of the material you'll be sampling, but it may not be the most efficient plan available. Knowing the AOQL of the plan will give you confidence in the amount of protection the plan affords you and your customers (refer to Figure 7-12). But it may not be the most cost-effective plan to use, as we saw when comparing the two sampling plans in Figure 7-15.

Sometimes, such as when new vendors, new processes, or new products are coming on-stream, you have to sample material without knowing the percent defective of the process. You can quickly estimate incoming lot quality by developing a p-chart and using the sample percent defective for the points on the chart. These charts are discussed in Module 6.

When you use a p-chart to estimate the quality level of incoming material, keep in mind that some of the variation seen in the chart may be due to assignable causes present during production of the material. This non-normal variation in quality may not be detected by your p-chart. The reason for this is that you may not be sampling the material in the sequence in which it was produced. The production sequence may have been mixed up after the material was produced or after it arrived at the sampling point. When you select samples from such lots, it is not likely that all the items selected were produced at the same time. If assignable causes were present, they may have been acting on only part of the items selected in the sample. For this reason, variation in quality due to assignable causes may be mixed in with sampling variation, and consequently the estimated percent defective will tend to be high.

SUMMARY

Some businesses find that the production and delivery of a service is not yet perfect. That is to say, defects occur at some rate in one or more of the operations of the process. As long as defects occur, there will be a need to use sampling to monitor and control the level of those defects.

Sampling can be performed in an unorganized manner, or it can be performed in a planned, efficient, cost-effective manner in which the risks are known. If we use acceptance sampling, the risks are known and the information derived is useful to management. Whether you choose to use sampling plans from standard tables or to develop your own plans, the material covered in this module will enable you to analyze and understand sampling plans and thus make better use of them to benefit your customers and your company.

Delivering a service that meets the wants, needs, and expectations of your customer should always be your goal. Many items or characteristics are necessary to produce and deliver your service. These items may be forms or documents used internally in the production of the service; they may also be an integral part of the service, such as periodic account status reports or a product sold as part of the service.

The process of delivering your service and the things used in the process will always be subject to variation. Sometimes this variation will cause defects to occur in your service. While you work continuously to improve the quality of your service, you may not completely eliminate these defects for a long time. In the meantime, whenever items are produced and handled in quantities, sampling plans can be used to monitor and maintain quality.

Acceptance sampling is used with a given lot of material in order to decide whether it is acceptable. The planning and effort you put forth to obtain samples will determine the quality of your decision to accept or reject the material. If the sample is biased in any way, the quality of that decision may be lessened.

A sampling plan can be a reasonable alternative to 100% inspection. Properly selected or designed sampling plans can be very useful in controlling quality by maintaining a limit on the average quality of material after it has been submitted to the plan.

Sometimes, you may need to estimate the quality level of material but you don't know the material's sequence of production. In this case, you could not use a p-chart to control the quality level. A sampling plan, however, would be a satisfactory statistical technique to use instead. By using a sampling plan, you can control the outgoing quality of material.

Sampling plans may be selected from published tables or they may be custom-designed to suit a particular need. Before selecting a sampling plan or designing one, you should consider and agree on the acceptable quality level (AQL) and the lot tolerance percent defective (LTPD).

Every sampling plan has parameters that determine how it will operate. The essential parameters are the sample size (n) and the acceptance number (c). These two parameters define the capability and limitations of the sampling plan. When designing a sampling plan, the sample size and the acceptance number can be adjusted. You can use these two parameters to develop the plan's operating characteristic (OC) curve. This curve will show you the risks associated with the plan.

Two types of risk are associated with sampling plans: the producer's risk (for the AQL) and the consumer's risk (for the LTPD). These risks often will be adjusted based on the economics of obtaining samples and inspecting or testing them. The amount of inspecting or testing involved in operating a given sampling plan is, in turn, determined by the capability of the process used to produce the units being inspected.

In this module we have learned how to maintain the average outgoing quality at an agreed-upon level. We can determine the average outgoing quality (AOQ) and the average outgoing quality limit (AOQL) for each sampling plan. We use this information when comparing the cost-effectiveness of several plans.

Costs to correct defects by replacement or repair after delivery to the customer, along with the cost of customer ill will, etc., must be considered when determining the acceptable quality level and the lot tolerance percent defective.

Remember, several things must be considered when you select a sampling plan. Consider whether you are selecting a sampling plan from standard tables or are designing your own plan to suit your needs. Other things to consider are:

1. The AQL, the acceptable quality level that will be accepted 95% of the time.
2. The LTPD, the maximum percent defective you are willing to accept 10% of the time.
3. The sample size.
4. The acceptance number.
5. The quality (percent defective) of the lots being submitted.
6. The desired AOQL, the average outgoing quality limit.

PRACTICE PROBLEMS: SAMPLING PLANS

Work through the following practice problems using the statistical techniques you learned in Module 7. The solutions to these problems can be found in the "Solutions" section beginning on page 272.

Problem 7-1.

Susan, the manager of a billing department, has determined that the average percent defective for documents generated in her department is .25% defective. This is the equivalent of 25 errors per 10,000 documents. Susan is willing to accept 25 errors per 10,000 documents leaving the department. If the error rate should rise as high as 250 errors per 10,000 documents, Susan wants assurance that the documents will be rejected and corrected before leaving the department.

Approximately 10,000 documents per day are processed through the department. They can be conveniently grouped in lots of 1,000 documents. Susan is trying to decide between two sampling plans for inspecting documents before they leave the department. One plan, called Plan C, calls for a sample size of 150 and an acceptance number of zero. The other plan, called Plan D, calls for the same sample size (150) but the acceptance number is 1.

Construct OC curves for the two sampling plans. Using the OC curves, evaluate their effectiveness based on their probabilities of acceptance at various percents defective. Which of the two plans is better for meeting Susan's goal?

Problem 7-2.

Construct the average outgoing quality curves for the two sampling plans in Problem 7-1. Estimate the average outgoing quality limit (AOQL) for each sampling plan. Which sampling plan has the lower AOQL? Do these estimates change your answer to Problem 7-1?

Problem 7-3.

Compare the total inspection required when using each of the two sampling plans in Problem 7-1. Assume the process average percent defective is .25%. The lot size is 1,000 documents for both sampling plans. Which sampling plan requires less total inspection? Using the information gained in Problems 7-1 and 7-2, which sampling plan will most efficiently meet Susan's operating goals in Problem 7-1?

Problem 7-4.

John is responsible for various merchandise that comes into a central warehouse and is distributed to outlet stores in the area. Towels are received from the supplier in boxes containing 500 towels per box and 10,000 towels per shipment. The supplier has agreed that .25% is an acceptable quality level.

If John decided to use acceptance sampling as a method of maintaining the agreed-upon quality, what sampling plan would you recommend he use? Refer to Table 7-1 and Table 7-2 for your selection of the plan, and assume John would use a normal level of inspection (General inspection level II).

MODULE 8
Systems Capability

NEW TERMS IN MODULE 8
(in order of appearance)

operation capability
system capability
probability plot
d_2
upper limit for individuals
lower limit for individuals
upper specification limit
lower specification limit

normal probability paper
estimated accumulated frequency (EAF)
plot points
line of best fit
capability index (C_p)
capability index (C_{pk})

INTRODUCTION

A process capability study predicts how well a process will do. But to make such a study, the process must be stable. A stable process, you will recall, is one in which no assignable causes are present; there is only inherent variation.

From Figure 8-1, we can see why we cannot do a capability study on a process that is not stable. Points 4, 8, 9, 10, 11, and 12 are outside the control limits. Every point outside the control limits is evidence that the overall average ($\overline{\overline{X}}$) has changed or shifted. When a process is not stable, the overall average for the process is moving around; it's not fixed in one place. Based on the data in this example, we cannot possibly predict where the overall mean will be in the future. Therefore, a capability study using this data would be meaningless.

In contrast, Figure 8-2 shows a process where the averages are stable. Since the averages are all inside the control limits, we know that most of the assignable causes have been removed. Once a process is stable, and we continue to keep it stable, we are then able to say that the overall mean is constant. Now we can do a process capability study and predict the percentage of services or parts of a service that will be inside or outside the specifications.

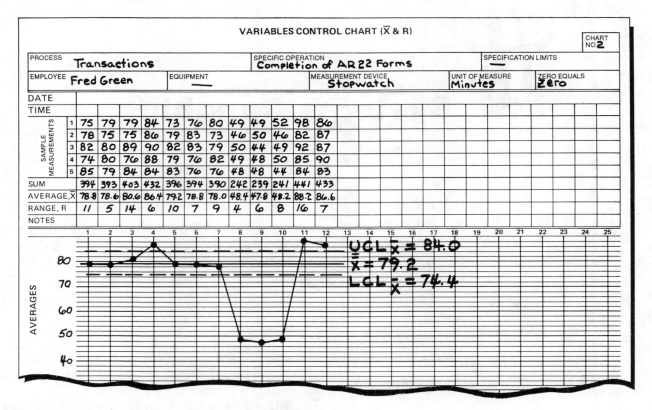

Figure 8-1. Average and range chart showing unstable process.

WHAT IS MEANT BY CAPABILITY

Capability describes how well a process meets its specified requirements. If a process meets requirements all the time, it is said to be capable. Sometimes, a process is stable but not capable. It is free of assignable causes yet still produces a service or part of a service that does not meet specifications. When this happens, it means the quality problem is an inherent part of the process. The process must be changed if it is to be capable.

In this module, we address two kinds of capability. One is *operation capability*, sometimes called machine capability. The other is *system capability*, or process capability. Operation capability is the short-term capability of an operation or part of the system. System capability is the long-term capability of the entire system.

You will be concerned with short-term capability mainly during start-up of a new operation or process. After a process is in place, you'll be

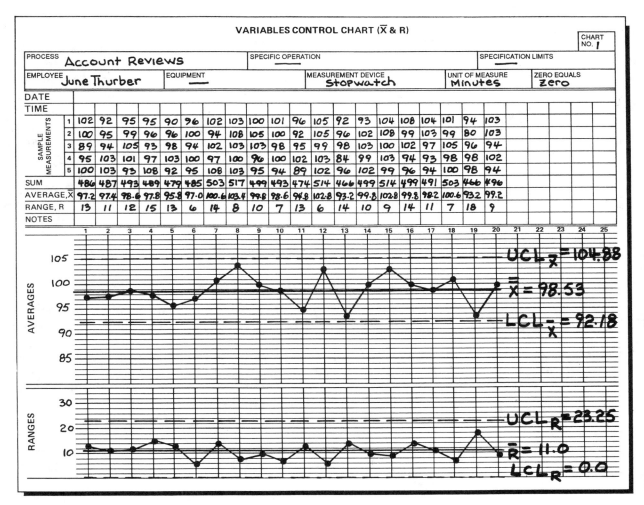

Figure 8-2. Average and range chart showing stable process.

concerned with the capability of the whole process. Sometimes, however, you'll need to make short-term capability studies to determine the effects of a change in materials, methods, or equipment.

In this module, you will learn how to measure an operation or process and compare these measurements with established specifications. By doing this, you can measure the capability of the process. Two tools are helpful for measuring the capability of operations and processes. These tools are the average and range chart and the *probability plot*. To obtain the best estimate of short-term (operation) capability or long-term (system) capability, these tools should be used together.

DEFINING THE SYSTEM

Before performing any capability study, you must first know what elements of your system you want to study. One way to know this is by defining your system. As you already know, a system for developing or delivering a service is usually made up of many elements or operations. To define your system, you must identify all the elements. Then determine those that are most critical to its performance. A process flow chart can help you do this by clarifying the sequence of operations that produce your service. After you have identified the critical operations, you must quantify them by establishing specifications for each one. This is done by specifying a value for each operation along with an allowable variation about that value.

The techniques for measuring short- and long-term capability are almost the same. The difference lies in how the measurements are obtained. We will discuss short-term capability first.

OPERATION (SHORT-TERM) CAPABILITY

An operation is generally considered a part of the total process or system. The entire process is subject to variation from such sources as employees, the methods they use, materials used in the process, the surrounding environment, and the machines or equipment used to perform various operations. (See the fishbone diagram in Module 1, Figure 1-4).

An operation capability study is intended to show only the variation caused by that operation and not the variation caused by other parts of the system: the environment, the incoming materials, the methods, or the employees. Although you cannot completely eliminate the effects of these other factors, you can minimize them by collecting measurements over as short a period of time as practical. When you do this, you can make the best estimate of operation capability.

An operation capability study is concerned with one element of a service at a time. One purpose of the study is to estimate the average of an element and compare it with specifications to see if it is well-centered between the specifications. A second purpose is to estimate how the operation value is clustered around the average. As we saw in Module 1, this pattern graphically shows the process spread. The process spread is measured in terms of the standard deviation of the operation. Standard deviation is simply a number we can use to describe the process spread.

AVERAGE AND RANGE CHARTS

When developing an average and range chart, we measure an operation and use these measurements to estimate the operation's stability. We need to know how stable the operation is to estimate its capability accurately. Our estimate of capability will be based on statistics obtained from the operation when all conditions are normal or stable.

The best way to determine stability is to make an average and range chart and look for the presence of assignable causes. If the chart shows points out of control, we have a signal that some assignable cause is distorting the normal distribution of the operation. We must eliminate assignable causes whenever possible. If we cannot do this, we will be less able to make an accurate estimate of operation capability.

We can also use a frequency histogram to roughly estimate an operation's capability. As you saw in Module 4, a frequency histogram can be used to estimate the average of a measurement and its spread around the average. The capability of an operation can be estimated by placing the specifications on the histogram. The histogram is useful if you need a quick check and are sure that no assignable causes are present, but the average and range chart is more precise and reliable for predicting operation capability.

When all assignable causes are identified and eliminated, you can say an operation is stable and predictable. Once your operation is stable, you can proceed with your capability study to see if your operation is within specification. A typical operation capability study is illustrated in Figures 8-3 and 8-4. This study measures the time it takes customers to complete bank transactions at a drive-through window. Figure 8-3 shows the control chart developed during the capability study.

In this study, measurements were converted into seconds to simplify the job of calculating the averages and the ranges. The measurements recorded were those of 125 consecutive customers. In this way, the measurements were taken in as short a time as possible, and variation due to causes other than the conditions prevailing at the time was kept down.

After all the measurements were recorded, we calculated the average range ($\bar{R}$), the upper control limit for the range (UCL_R), the average of the sample averages ($\bar{\bar{X}}$), and the upper and lower control limits for the average ($UCL_{\bar{X}}$ and $LCL_{\bar{X}}$, respectively). As with any average and range chart, we developed the range chart first and determined that there were no out-of-control points on that chart before proceeding with the average chart.

Fortunately there were no points outside the control limits (see Figure 8-3), but this is not always the case. Often it is very difficult to find and eliminate assignable causes. But it is often worthwhile to proceed with a capability study even if you cannot eliminate all the assignable causes. When you do this, however, you should be very cautious in making your capability estimate. Keep in mind that assignable causes tend to distort the normal frequency distribution curve. When this happens, you are less able to predict the capability of an operation accurately.

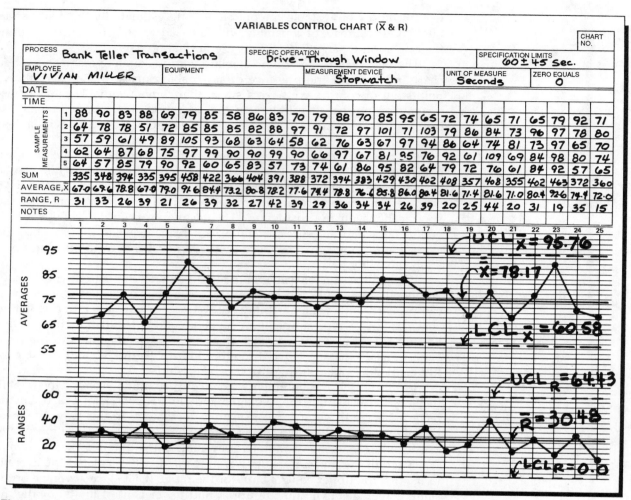

Figure 8-3. Average and range chart developed during operation capability study.

LIMITS FOR INDIVIDUALS

After you have established the average and range chart with control limits, you must estimate the upper and lower limits for individual items. You can perform this simple calculation using the form on the back of the average and range chart. The formulas are supplied; you just put in the numbers. Use the section of the form headed "LIMITS FOR INDIVIDUALS." See Figure 8-4.

Figure 8-4. Back of average and range chart showing "Limits for Individuals," upper right.

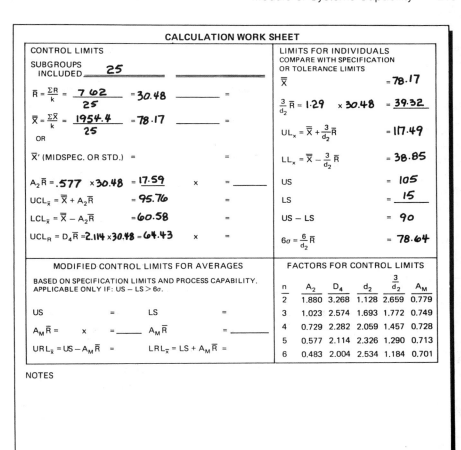

To help you understand and use this form, we will look at a number of terms. Some of these terms are new and some you have already learned.

$\overline{\overline{X}}$: Overall average or mean of the process or operation.

$\overline{R}$: Average range of the operation.

σ: "Sigma," the standard deviation of the process or operation. One of the divisions of the distribution around the average of the normal curve. (See Module 1, Figure 1-7.)

d_2: A factor used in combination with the average range ($\overline{R}$) to determine the standard deviation (σ). (See d_2 under "FACTORS FOR CONTROL LIMITS," Figure 8-4.) 1σ equals $\overline{R}$ divided by d_2; 3σ equals $3\overline{R}$ divided by d_2.

UL_X: *Upper limit for individuals;* the estimated largest individual value to be produced by the operation under study. This is not to be confused with the upper control limit ($UCL_{\overline{X}}$), which is the limit for averages. UL_X equals $\overline{\overline{X}}$ plus ($3\overline{R}$ divided by d_2).

LL_X: *Lower limit for individuals;* the estimated smallest value to be produced by the operation under study. LL_X equals $\overline{\overline{X}}$ minus ($3\overline{R}$ divided by d_2).

USL: *Upper specification limit.*

LSL: *Lower specification limit.* USL minus LSL equals specification limit, or tolerance.

6σ: Formula for process spread (operation capability). $6\overline{R}$ divided by d_2 equals 6σ (operation capability).

If the upper limit for individuals (UL_X) is equal to or less than the upper specification limit (USL), and if the lower limit for individuals (LL_X) is equal to or greater than the lower specification limit (LSL), you can say that the operation is capable of meeting the specification. Remember, this is true only if the measurements show the operation to be stable and normal. Figure 8-5 illustrates a stable condition: (1) the process spread

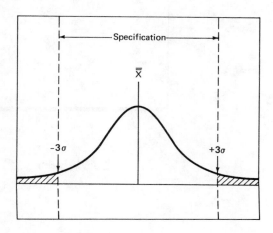

Figure 8-5. Process spread is normal; coincides with specification limits.

Figure 8-6. Process spread is greater than specification limits.

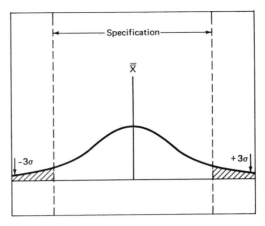

and the specification limits are the same and (2) they coincide. Most of the items (99.73%) produced in an operation like this will be within specification. However, you can see that if the overall average ($\overline{\overline{X}}$) changes in such an operation, items will be produced outside the specification.

In a case where the process spread is greater than the specification limits, the operation must be declared not capable of meeting the specification. See Figure 8-6.

In a case where the process spread is less than the specification spread, but the upper limit for individuals (UL_X) or the lower limit for individuals (LL_X) is outside the specification limits, an adjustment is required in the process average or process mean ($\overline{\overline{X}}$). In Figure 8-7, the curve has the same spread as in Figure 8-5, but the process mean ($\overline{\overline{X}}$) is on the "high" side of the center. As a result, the lower 3σ limit is well within specification, but the upper 3σ limit is well above the specification limit. This process could be brought within the specification limits if an adjustment is made to shift the process mean to a lower value.

Figure 8-7. Process mean is high; upper limit is above upper specification limit.

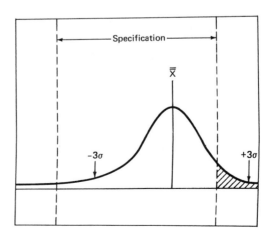

Figure 8-8. An ideal condition; the process is centered on the specification and the process spread is less than the specification.

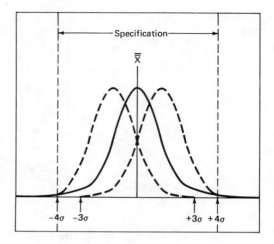

A much better situation is shown in Figure 8-8. The process is centered on the specification and the process spread is such that 8σ is equal to the specification. In this case, the operation can change somewhat and the items being produced will still be within specification.

This is your goal. The measurement created by the operation being studied must be centered on the specification and the process spread must be as small as possible. Keep in mind that the capability of one operation is only a part of the total process or system. Over a period of time, other sources of variation will affect the output of this operation. These other sources will cause the spread to be greater than the estimated process spread created by the operation alone.

When you use an average and range chart to predict process or operation capability, you have the advantage of seeing how stable your operation is because the chart can show you whether or not assignable causes were at work when the measurements were taken. However, you can't always be sure about the shape of the distribution of the original measurements. If the process is stable—no points outside the control limits—the distribution of the sample averages plotted on an average and range chart is nearly always normal, regardless of whether the original individual measurements formed a normal frequency distribution. We assumed the frequency distribution was normal, as shown in Figure 8-5, when we compared the operation spread to the specification. When we made calculations (Figure 8-4) to estimate the upper and lower limits for individual items and to determine the operation spread, we assumed here, too, that the distribution was normal.

The charts shown in Figures 8-3 and 8-4 are typical of an operation capability study. The measurements were taken on a continuous basis to minimize the variation due to outside sources. As a result, the variation shown on the chart is due mostly to the operation that created the measurement under study, namely, the time to complete a transaction at a drive-through window of a bank. No points were outside the control limits, so the operation was considered stable enough to continue the study.

In our example, the overall average ($\bar{\bar{X}}$) is 78.17. This value is entered in the "LIMITS FOR INDIVIDUALS" block on the back of the control chart. See Figure 8-4. The estimated spread about that average is calculated by multiplying a factor called $3/d_2$ by the average range. This factor is found in the "FACTORS FOR CONTROL LIMITS" table on the back of the control chart.

The column headed "n" in this table lists the subgroup sizes. Our average and range chart was developed using a subgroup size of 5, so we move from the 5 in the n column to the column headed $3/d_2$. The corresponding number in this column is 1.290. We then multiply this number by the average range (30.48) to get 39.32.

We obtain the upper limit for individuals (UL_X) by adding this number (39.32) to the average ($\bar{\bar{X}}$). We obtain the lower limit for individuals by subtracting the same number (39.32) from the average ($\bar{\bar{X}}$). Here, the UL_X is 117.49 and the LL_X is 38.85.

The estimated total spread of the operation is equal to "six sigma" (6σ), which is six times the range divided by the factor d_2 (2.326). Here, our answer is 78.64 seconds. This is the estimated capability of our operation. Keep in mind, however, that we are assuming the frequency distribution of our original measurements is normal. To verify the accuracy of our capability estimate, we can develop a probability plot of the measurements and check our frequency distribution.

THE PROBABILITY PLOT

A simple way to test the accuracy of estimates made from control chart measurements is to use a probability plot to estimate the shape of the distribution formed by those measurements.

When developing a probability plot, we take the measurements from our average and range chart and plot them in such a way that we can estimate how well they fit a normal curve. We use this technique with the average and range chart method to make the best estimate of operation capability.

This method uses a special graph paper called *normal probability paper*. See Figure 8-9. This form of graph paper was developed through the efforts of many people in the field of quality control. Each step in the development of this paper has helped to simplify its use. If you can't find the form for the normal probability chart used in this book, you can use some other form or make copies of the form used here.

Figure 8-9. Normal probability chart. (Form developed by Howard Butler.)

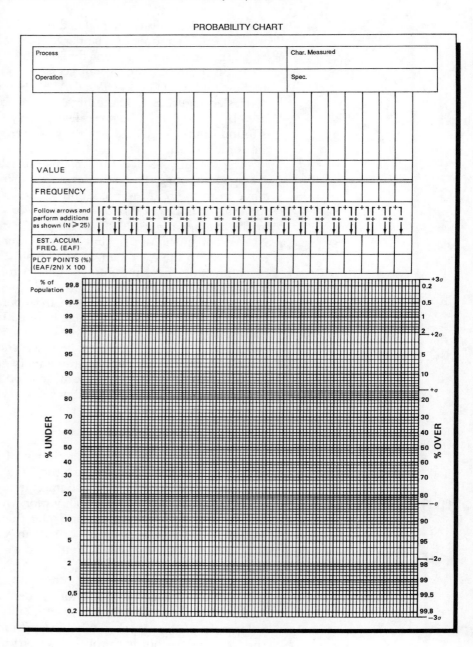

Using normal probability paper, we will be able to show the following:

- the center of the frequency distribution curve (where most of the measurements are)
- The spread of the frequency distribution curve (how the measurements cluster about the middle)
- the general shape of the frequency distribution curve
- the percentage of measurements beyond the specification limits

Now you will learn how to construct a probability plot using normal probability paper.

Step 1. Gather the information and fill in the heading.

We will use the measurements from the first 20 subgroups on the chart shown in Figure 8-3—a total of 100 measurements. You can estimate the normality of a process with fewer measurements. Fifty is enough, and, in fact, many estimates of normality and capability are made with as few as 25 measurements. When you use fewer than 50 measurements, however, you run a high risk that your estimate will be wrong. On the other hand, it's hardly worth the trouble to make more than 100 measurements. Since the information is available, you might as well take advantage of the greater degree of accuracy provided by the larger sample of 100. You could use all 125 measurements, but the calculations for 100 measurements are much easier to do.

The chart we will use to make our probability plot is shown in Figure 8-9. Enter the information concerning this study in the heading of the chart. This includes the operation name, the element measured, and the specifications. See Figure 8-10.

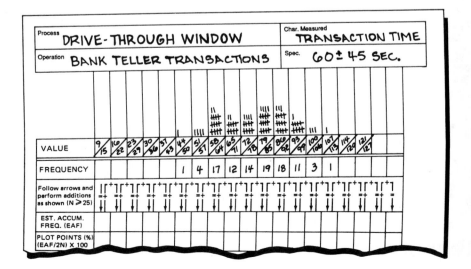

Figure 8-10. Normal probability chart with heading information, tallies, values, and frequencies.

Step 2. Make a tally of the measurements.

Immediately under the heading is a section in which to make a frequency tally of the data. (This is also shown in Figure 8-10.) To make this tally, we will use the method described in Module 4. First, set a scale that will give you about 10 divisions. In this example, an interval of seven works well. In the "value" row, the first block or cell should contain the lower specification limit (which is 15). The block interval for that value is 9/15. This means that the values 9, 10, 11, 12, 13, 14, and 15 will be tallied in that section. The next block in the value row will contain the numbers 16/22, the next 23/29, and so on, up to the last block, which will contain 121/127.

As you tally each measurement, enter each tally mark in the appropriate block. For example, make a mark in the "72/78" block for each measurement of 72, 73, 74, 75, 76, 77, or 78. When all measurements are tallied, record the number of tallies in each block in the corresponding blocks in the "FREQUENCY" row—located directly below the "VALUE" row.

Step 3. Find the estimated accumulated frequencies.

To plot points on the graph, you must transform the frequency of each tallied measurement value, to an *estimated accumulated frequency (EAF)*. You must then convert these estimated accumulated frequencies into percentages of the total sample. This may sound confusing at first, but with use, you will find it quite simple. You can use the percentages to estimate how normally the measurements are distributed around the average.

This is how to calculate estimated accumulated frequencies (EAFs), as shown in Figure 8-11.

Begin in the frequency row. Transfer the first number on the left to the block directly below it in the EAF row. In this case, the first number is 1.

Take the EAF number you just entered (1), follow the arrow located above the block it's in, up past the plus sign, to the "FREQUENCY" row. Add the first frequency (1). Then follow the arrow to the right, past the next plus sign, to the second frequency (4), and add that number. Follow the arrow down, past the equal sign, and enter your total (6) in the second EAF block. In our example, $1+1+4=6$.

To find the third EAF value, start with the second EAF value (6), follow the arrow upward, and add the next two frequency counts (4 and 17) to obtain 27. This is your third EAF value. In our example, $6+4+17=27$.

Follow the arrows and add the numbers until you have added all the frequency counts. The last arrow in our example starts with an EAF of 199. Follow the arrow up to the last frequency count (1) and add this number to the 199. Then, move to the next frequency block (which is

Figure 8-11. Normal probability chart showing estimated accumulated frequencies (EAFs).

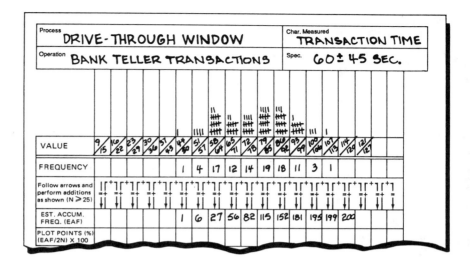

blank) and continue down to the EAF block and record your total of 200. At this point, we can check our addition simply by comparing the largest EAF to the sample size. This value should always be twice the sample size. In our example, the largest EAF value is 200 and the sample size is 100.

Step 4. Plot the points.

Now you can use the estimated frequencies to plot the points on the probability graph. To obtain the points to be plotted on the graph paper, you must convert the EAF values to percentages of the total sample that formed the distribution. (See "PLOT POINTS" row in Figure 8-12.)

To convert the EAF frequencies into percentages, use the formula in the "PLOT POINTS" row. Divide each EAF by twice the sample size (2N) and multiply the result by 100.

>Plot points (%) equals EAF divided by 2N, all times 100.

In our example, N equal 100; so 2N equals 200. The first EAF is 1. Therefore, the first *plot point* is $(1/200) \times 100 = 0.5$. The second point is $(6/200) \times 100 = 3$; the third is $(27/200) \times 100 = 13.5$; and so on, as shown in Figure 8-12.

Now you are ready to plot the points on the graph portion of the chart. On the left-hand edge of the graph, which is marked "% UNDER," find the number that matches the number in the "PLOT POINTS" box. On this chart, the number in the first box in the "PLOT POINTS" is .5,

Figure 8-12. Normal probability chart showing plot points.

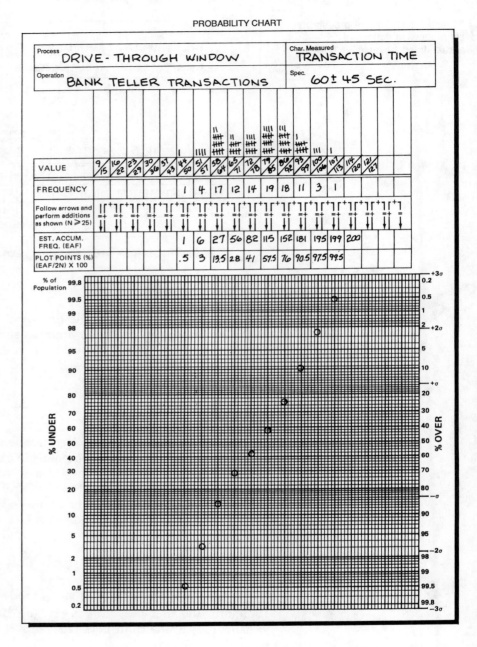

so locate the number .5, or 0.5, on the "% UNDER" scale. Follow the horizontal line that corresponds to 0.5 until you reach the heavy vertical line that runs up to the first "PLOT POINTS" box. Place a point where these two lines meet and draw a small circle around the point so you can see it more clearly. See Figure 8-12. The number in the next "PLOT POINT" box is 3, so find 3 on the "% UNDER" scale. Follow the horizontal line across until you reach the line that runs up to the "PLOT POINT" box containing the number 3, and plot the point. Continue doing this until you have plotted the point for the "PLOT POINT" box containing the number 99.5. See Figure 8-12.

Step 5. Draw the Line of Best Fit.

Now draw a straight line that best fits the plotted points. This line is called the *line of best fit*. If the measurements used in this study are distributed normally around the average, the points will fall in a straight line on the graph. Any trend away from a straight line is a sign that the normal distribution is distorted. A line has been fitted to the plot points in Figure 8-13.

At this point, you can judge the accuracy of your estimate of the process spread, which you made using the average and range chart. If the plot points fall on or nearly on the line of best fit, you can be confident about the accuracy of your estimate of the mean and standard deviation based on the average and range chart. If the plot points do not fall approximately on the line, you should have less confidence in your estimate.

Step 6. Estimate the process spread.

To estimate your process spread, extend the line of best fit beyond the points you plotted on the graph until it intersects the horizontal lines at the top and bottom of the graph. You will use the values represented by these two intersection points to estimate the process spread.

In Figure 8-13, the line of best fit crosses the bottom of the graph at a point that is directly under the values 37/43 in the value row. This means the lowest value from this operation will be at the low end of these values—in this case, the value is 37.

The line of best fit intersects the top of the graph at a point that is directly under the values 114/120 in the value row. This approximately equals a value of 117. We can expect, then, that the largest value from our operation will be 117, or 117 seconds, for the longest transaction.

The estimated process spread of our operation, then, is from 37 to 117. This is equal to six sigma (6σ), or six standard deviations of the operation. Under normal operating conditions, we can estimate that 99.73% of all our transactions will be within a range of 37 to 117 seconds. This is the capability of our operation when no assignable causes are present.

We can also use our probability plot to estimate the process average of our operation. The process average is represented by the value of the point where the line of best fit intersects the horizontal "50%" line on the

Figure 8-13. Normal probability chart showing line of best fit.

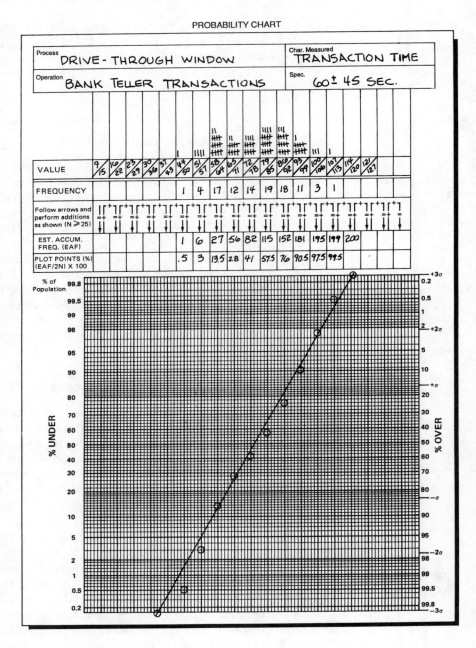

graph. By moving across our 50% line to the line of best fit and then moving straight up to the value row from this intersection point, we find that the value block represented is 72/78. An imaginary vertical line drawn up from the intersection point enters the 72/78 value block at approximately the 77 value. This is our estimate of the overall average of the transaction times.

The limits for individuals and the overall average as calculated using the average and the range determined in developing the control chart are slightly different from those obtained from the probability plot. The estimated smallest value is 38.85 and the largest value is 117.49. The overall average is 78.17.

The estimates for the average are very close. The estimates for the limits for individuals are such that the process spread is greater when using the probability plot than when using the average and range values. This difference occurs because the probability plot shows that the measurements do not all fall on a straight line. If the underlying distribution of the measurements is distorted, the estimates made using the probability plot are more likely to be closer to the true process spread than those made from the average and range chart. It must be kept in mind at all times that SPC techniques such as these use relatively small samples to estimate process variations. Since the estimates are made using samples, they are subject to error and may vary from the true values.

ESTIMATING THE PROPORTION OF MEASUREMENTS OUT OF SPECIFICATION

You can use the graph obtained by the probability plot method to estimate the proportion of measurements that will be out of specification. The specification limits can be drawn on the graph, as shown in Figure 8-14. If the line of best fit crosses a specification limit line before it crosses the top or bottom edge of the graph, you can estimate the percentage of measurements out of specification by reading the percentage above the upper specification limit in the vertical scale on the right-hand side of the graph ("% OVER"), opposite the place where the line of best fit crosses the upper specification line. You can estimate the percentage below the specification by reading the percentage on the left vertical scale ("% UNDER"), opposite the place where the line of best fit crosses the lower specification line. These estimates can be used even if the plot points do not fit well to a straight line, but the estimates won't be as good.

Figure 8-14 shows that our drive-through transactions operation is not capable of meeting the specification because:

- the operation is not centered on the specifications. The specification mean is 60 seconds and the process mean is 77 seconds.
- the upper limit for individuals (UL_X) is greater than the upper specification limit (USL), 117 seconds versus 105 seconds.

The line of best fit crosses the upper specification limit line at approximately the "2% OVER" line on the graph. This means that when the

Figure 8-14. Normal probability chart showing specification limits.

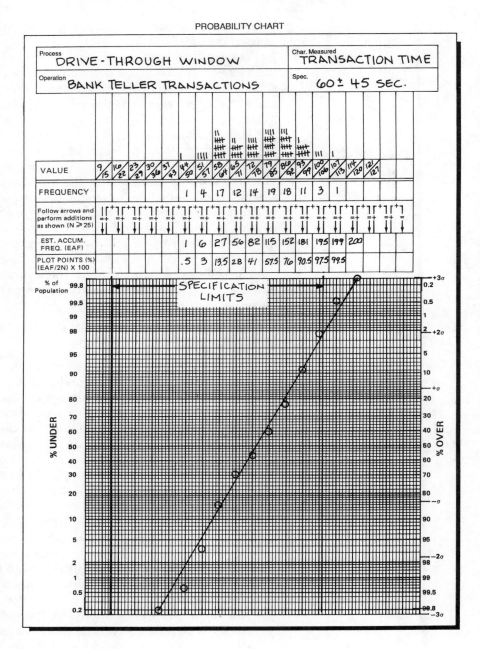

operation is free of assignable causes and the process mean is 77, approximately 2% of the transactions will be greater than the upper specification limit of 105 seconds.

SYSTEM CAPABILITY

Most of the time you will be concerned with operation (short-term) capability, but occasionally you will need to know the capability of a whole system as it relates to an operation (long-term capability.)

The basic difference between operation capability and system capability is that operation capability is concerned with the variation in measurements caused by only one set of conditions in an operation, such as the same equipment, operators, or environmental conditions. System capability is concerned with the variation caused by all the sources of variation: the changes in equipment, the different operators involved, the materials used, the methods employed, and the environment as it affects the service. These are things that change over time. Therefore, the study of system capability must be made over a longer period of time.

If you want to know the capability of a system, your measurements must include the chance variation from all sources. This includes the variation that is built into the process but which occurs only over a relatively long period of time. Changes in materials used, shifts in temperature, equipment wear, use of different personnel, and small changes in methods are all part of the system and should be included when you estimate system capability.

To make a system capability study, you must collect your measurements over a fairly long period of time. Once you have recorded your measurements, the analysis of these measurements is similar to the analysis you made for the operation capability study.

These are the steps in making a system capability study:

Develop an average and range chart using measurements taken over a fairly long period of time. In many operations, 30 days is long enough to obtain measurements that include variation from all sources. Your knowledge of the system, however, will be the most valuable factor in deciding the length of the time over which your system must be sampled.

Identify and note all assignable causes, especially those that make the range go out of control.

Make a probability plot of the measurements used to develop the average and range chart. This plot must be made using measurements covering the entire time span of the study.

Estimate the process spread from the probability plot and evaluate the system capability by comparing the process spread with the specification limits.

The average and range chart will give you a sense of the stability of your process. The range is a measure of the process spread at the time the sample was taken. Ranges are most often considered reflections of an operation's short-term capability, whereas averages are most often affected by variation from sources whose changes are long-term. Do not be surprised, then, to see the averages move up or down as new materials are used, as the weather changes, as equipment wears, or as other things happen that affect the people or the environment involved in the production of the service being studied.

Because the range tends to show only isolated, short-term variation and the average is affected by variation from all sources, you must estimate system capability using the individual measurements that make up the averages.

As with an operation capability study, the probability plot can help you estimate the shape of the distribution of measurements around the average. This is also the recommended method for estimating system capability. If the probability plot is a straight line, or nearly straight, you can use it to estimate the process average and process spread. (There are also many more complicated ways of estimating system capability, but we'll leave those to the statisticians and engineers.)

CAPABILITY INDEX

Once you have determined that your process or operation is stable (in statistical control) and distributed normally, you can evaluate its capability. Remember, however, that you must first calculate the process spread using an average and range chart and then analyze the shape of the distribution with a probability plot.

The capability of an operation or a system can be expressed as a number called the capability index. This is a handy way to talk about the capability of any operation or system, whatever the service may be. As you will see in what follows, you find the index number by comparing the process spread to the specification spread, and express this comparison in terms of the standard deviation.

One form of *capability index* (C_p) is simply the ratio of the specification spread, or tolerance, to the process spread, or six standard deviations (6σ) of the process. A C_p of 1.0 or greater means that the operation or system is capable of producing the service with a spread less than the tolerance.

This is the formula for finding the capability index:

C_p equals the tolerance divided by 6σ of the process.
C_p = tolerance/6σ of the process

A process with 6σ equal to the tolerance will have a C_p of 1.0.
The C_p for our bank transaction time operation is

$$C_p = \text{tolerance}/6\sigma = 90/78.64 = 1.14$$

Figure 8-15. Calculation of capability index.

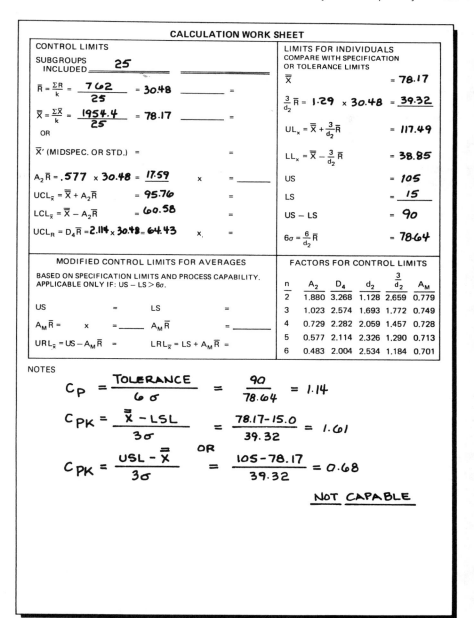

See Figure 8-15.

The C_p index does not tell the whole story, though. With this method, you could show a C_p larger than 1.0, which indicates that the spread is less than the specification tolerance and, therefore, capable. However, such a process could be producing all items outside the tolerance. The C_p index does not take into account where the process is centered with respect to the required tolerance. See Figure 8-16.

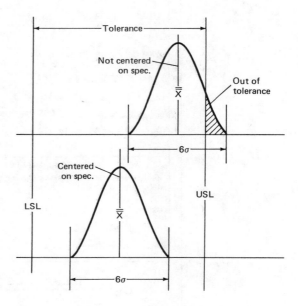

Figure 8-16. Two processes showing the same capability index (C_p). The upper curve is not centered on the specification; the lower curve is centered on the specification.

A second type of *capability index* (C_{pk}) is now used by many companies. This capability index will be a negative number if the process average is outside the specification tolerance. Any number less than 1.0 is an indication of a noncapable process or operation. Parts or services produced may fall outside the specification limits, although we may not find any outside the specification limits on a short-term basis.

The C_{pk} is calculated as the smaller of these two values:

The upper specification limit minus the process average, all divided by three standard deviations of the process, or

the process average minus the lower specification limit, all divided by three standard deviations of the process.

C_{pk} then, equals the smaller of:
(USL minus $\overline{\overline{X}}$) divided by 3σ, or ($\overline{\overline{X}}$ minus LSL) divided by 3σ.

The C_{pk} for our bank transaction time example is

$$C_{pk} = (\overline{\overline{X}} - LSL)/3\sigma = (78.17 - 15)/39.32 = 1.61$$

or

$$C_{pk} = (USL - \overline{\overline{X}})/3\sigma = (105 - 78.17)/39.32 = 0.68 \text{ (NOT CAPABLE)}$$

See Figure 8-15.

Remember that the process spread is 6σ, so half the spread, or 3σ, will fall on each side of the process average. See Figure 8-17.

SUMMARY

There are two types of capability analysis: operation, or short-term capability studies, and system, or long-term capability studies.

The statistical methods used in these two techniques are the same and are based on the six basic principles of statistical methods of quality control.

An operation capability study analyzes the variation in an element of a service caused by one operation or process of the system. This operation is generally only one part of the total system. A study of an operation covers a relatively short period of time.

A system capability study analyzes the variation in an element of a service caused by all sources of variation in the system. A system capability study analyzes the output of a system or operation over a fairly long period of time.

The estimate of capability of an operation or a process is developed using an average and range chart and a probability plot. The average and range chart enables you to determine the presence of any assignable causes that would indicate the operation or process was not stable. It also provides the overall average and the average range. These values are used to estimate the upper and lower limits for individual measurements and to estimate the standard deviation of the operation.

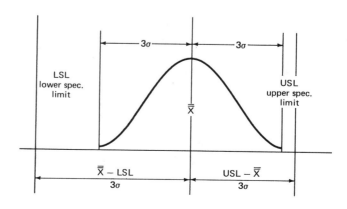

Figure 8-17. Two possible sources of C_{pk}. True C_{pk} is the smaller of these two values.

The probability plot is a method for judging how well the measurements fit the bell-shaped curve. The estimates made using the average and range chart assume that the measurements do fit the bell-shaped curve. The probability plot is used to estimate the standard deviation when a process capability study is being made. This is because the measurements are taken over a long period of time and because using the average range to estimate the standard deviation may result in a value that is smaller than the true value.

The capability of operations and systems can be described by the capability index. The capability index makes use of a number to describe the spread of the operation or system in relation to the specification tolerance and the relation of the process mean to the specification limits.

PRACTICE PROBLEM: SYSTEMS CAPABILITY

Problem 8-1

Work through the following practice problem using the statistical techniques you learned in Module 8. The solution to this problem can be found in the "Solutions" section beginning on page 279.

A hotel manager evaluates the quality of room cleaning in a hotel by inspecting some rooms each day. A system of demerits has been devised that results in a number for each room inspected. The system has been designed to measure the capability of the housekeeping system to meet a target of a maximum of 25 demerits for a room. Five rooms are inspected each day. Listed are the demerits obtained for 20 days.

Is the housekeeping system capable of meeting the demerit target of 25 maximum for a room? What is the capability index for this system?

Date	Room #1	Room #2	Room #3	Room #4	Room #5
8/1	19	7	17	21	14
8/2	24	16	12	15	20
8/3	20	20	16	20	16
8/4	18	21	25	21	16
8/5	18	16	21	19	19
8/6	20	16	20	25	22
8/7	9	21	23	15	11
8/8	23	11	24	15	8
8/9	15	18	26	17	16
8/10	14	10	13	11	19
8/11	16	14	28	10	15
8/12	4	11	18	16	23
8/13	12	20	11	18	14
8/14	21	24	20	21	23
8/15	20	20	26	14	9
8/16	22	23	16	18	13
8/17	18	19	27	15	23
8/18	18	21	6	20	7
8/19	20	17	15	14	15
8/20	25	17	21	28	15

Number of Demerits

MODULE 9
Putting It All Together

Throughout this book, we have been describing tools for controlling and improving the processes that provide your services. But which tools do you use when and where? And how do you present your work in a convincing manner? In this module, we provide a framework in which to use the tools for improving and controlling your processes as well as communicating the results.

It is not enough just to find the solution to a problem. We need to tell people how we did it, how much the solution will cost, and how effective it will be. Because we have to sell our ideas to others, it is essential that we communicate our results in a way that is likely to gain acceptance. Often, we need management's approval for a change in the process. Or we may need another department to interact with us. This is why a framework for solving problems and communicating your results is helpful.

As you read this module, think of yourself as working with a team to present the team's work to an audience of managers, staff, or other teams. Also think of this framework as a simple, straightforward way to write up your team's problem-solving activity. In both situations, we want you to see that problem solving follows a clear, simple series of steps. Once you and others become familiar with these steps and the tools and techniques of this book, you will find it easy to communicate what you have done to solve problems.

There are different names for this framework. Many Japanese problem-solving teams use the term "QC Story." Florida Power & Light calls theirs the "QI Story." Some divisions in the automotive industry refer to the "Corrective Action Process." By whatever name, the framework becomes a standard method of solving problems in the company. Where everyone in the company uses the same method, it becomes a common language. This greatly enhances communication.

This framework, like good problem solving, follows a series of steps:

1. Introduction
2. Selection of problem
3. Investigation and analysis
4. Countermeasures
5. Results

6. Standardization or adjustment
7. Conclusions and plans

Let's look briefly at each of these steps and which tools apply.

FRAMEWORK FOR SOLVING PROBLEMS

Step 1. Introduction

Ideally, you will work with a team of fellow employees to make any improvements or to establish ongoing controls. When you and your team present the results of your problem solving, the audience needs to know who you are, who was in the team, what department you are from, and the time period involved. They will also want a brief outline of the process. This is background information similar to the kind of information you wrote on the various charts you've worked in this book. It gives your audience perspective.

Step 2. Selection of problem.

You probably notice things about your job, work area, or process that can be improved or that need monitoring on a regular basis. To have a clear statement of the situation, you must first do a little background work.

One thing to do is talk to your customers. These people include the final customer: the customer at the drive-through window at the bank, a patient having X-rays taken, or a traveler coming into the hotel room. But the "customer" is also the person who is next in line in a process: the server is a "customer" of the salad chef, and a trainer is a "customer" of the graphic artist who makes up overhead transparencies for the classroom. Find out what your customers think, what they want, and what they need from you.

Another thing to do is take a close look at your work activities and whatever quality standards apply to the work.

Ask management what they think.

A good tool to use for determining possible places for improvement or control is *brainstorming*. Another tool to consider is the *process flow chart*. But once you have identified several problems, you will have to make a selection. If you have, or can get, appropriate numerical data on the relative sizes of the problems, *Pareto analysis* should help in making the selection.

Step 3. Investigation and analysis.

There are two stages here. First, study the current situation without trying to develop solutions. This is when you "play Sherlock Holmes": Go to the scene of the event (the problem), and look carefully at the situation to detect and decipher clues. Do not jump to solutions. Instead, learn all you can. Talk to the people involved; review records about the process; watch the process first-hand. The team may find either the *process flow chart* or the *process cause and effect diagram* useful in its investigation of what is currently going on.

In the second stage, you dig in. An excellent tool for your team at this point is a *brainstorm* followed by a *cause and effect diagram*. Use these tools to identify all possible causes your team can think of as well as the relationships of one potential cause to another. Another way to approach the problem is to *storyboard* for possible causes and their relationships.

Other tools are data-based. The *frequency histogram* can tell you about the distribution of data, where they are centered, and how well the process meets specifications. *Control charts* tell you whether or not the process is in statistical control. *Process capability studies* tell you whether your data follow the normal, or bell-shaped, curve and what percent of the elements of the process being measured are within specifications. Other effective tools to use here are *checksheets* to organize the data and to help provide clues on where the problem occurs, *scatter diagrams* to help you identify relationships between variables, *storyboarding* to break the problem into components, *process flow charts* so your team can look at the overall process and think about where trouble is occurring. Perhaps constructing another *Pareto diagram* will focus your efforts more clearly.

Step 4. Countermeasures.

Once your team has determined what one or more factors are real causes, or at least which factors are very good candidates, then it's time to think about solutions. An effective tool to use here is the *storyboard*. The team can storyboard solutions or countermeasures to deal with the factors they have identified. (We call solutions countermeasures because they are "measures" taken to "counter" the causes of the problem.)

There may be any number of possible solutions. For example, in dealing with certain kinds of errors made in processing documents, your team may decide that a new module of training for new employees will stop the problem. If you have been working to improve customer goodwill, you may decide a new layout of your retail outlet will accomplish your goal.

But beware at this point! The countermeasures or solutions you have developed may do the whole job neatly and cleanly. Sometimes, however, they don't. They may only partially solve the problem. They may cause other problems or they may be too expensive. You need to check.

Step 5. Results.

Now you must check the countermeasures to verify their effectiveness in solving the problem, to see if they cause other problems, and to see if they are cost-effective. You can document changes to the process by comparing *frequency histograms* done before you began the project with ones developed at this stage. You can use *control charts* and *process capability studies* in a similar manner. A *Pareto diagram* will show if the same categories of problems are occurring with the same relative importance. Your hope is that the large category your team worked on is now only a small bar on the Pareto diagram, or even gone completely.

Step 6. Standardization or adjustment.

If your team is satisfied that your solutions have solved the problem, you must show what has been done to *keep* the solutions in place so that the problem never reoccurs. This is what we mean by standardization—developing and putting in place new standards so that the problem stays solved.

Sometimes, though, your team's first attempts do not fully accomplish your plans or goals. There are still too many errors in the documents. Or results of customer surveys, though better, are not where you want them. In this case, Step 6 means you must make an adjustment. Since you have not reached your goal, you need to modify or "adjust" your solution. Now is the time to stand back from the problem, rethink it, and possibly develop new plans. Ask yourself: What has our team tried? What have we accomplished so far? In what ways does our plan fall short? And most important, what do I and the team plan to do about it? Include these questions and their answers when you make your presentation.

Now let's see where the SPC tools and techniques may apply to standardizing or to adjustment. When you are standardizing the solution, the tools of statistical process control can help you monitor how well the solutions are working. For example, if you develop a module for training new employees in preventing certain types of mistakes on the documents they process, a *control chart* can monitor the effectiveness of this training. *Pareto analysis* of the document errors every week can tell whether the training is actually doing the intended job of preventing these errors. Sometimes because of the nature of the problem, you may need to make *frequency histograms* or do *capability studies* to check your solutions on a regular basis.

On the other hand, suppose the results have failed to show the improvement you had anticipated. Perhaps you implemented your solutions to improve customer goodwill, but still the surveys aren't good enough. Now you must adjust the original planned countermeasure. This may mean minor modifications or it may mean major changes. How do you do this? What tools and techniques do you use?

This is a good time to go back to the *storyboard* or *cause and effect diagram* your team developed showing possible causes of the problem. Were your solutions hitting the correct causes but inadequate to do the job? If so, see if you can strengthen the solutions. Perhaps you set aside certain causes or "bones" on the original cause and effect diagram as not being the true causes. Further analysis could show they are really important after all. (This is why it is important to keep the original storyboard or cause and effect diagram with all its bones.) Maybe you will need to have a *brainstorm* session and build a new *cause and effect diagram* or do another *storyboard*.

Adjustment, as you now should see, reopens the door to all the problem-solving tools and techniques. After another look at your original cause and effect diagram or storyboard, or development of new ones, you may have to collect more data, develop a new *Pareto diagram*, or use any of the other tools.

Step 7. Conclusions and plans.

In Step 7, your team has the opportunity to express any observations or conclusions you have reached during your problem-solving activity. What things might be improved further? What did you learn about working as a team or about problem solving?

This is also the time to present your plans. If you were not able to accomplish your original goal so that in Step 6 you had to make an adjustment, you need to describe your adjustment plan. If you were able to accomplish your original goals, you need to describe your new plans—what new project are you going to work on?

SUMMARY

The framework we have been describing is a model for taking corrective action to improve both your services and the systems that produce and deliver them to the customer. Activities for improvement are dynamic because we're not just living with the old problems—now we have the means through the tools and techniques described in this book to do something about our problems, to mend what is amiss and to make what is already good even better. Using the tools and techniques of quality in the context of the framework will help your company attract and keep customers, increase its market share, and become more profitable.

Solutions to Practice Problems

MODULE 2: BASIC QUALITY PROBLEM-SOLVING TOOLS

Problem 2-1a.

Whenever you are brainstorming, make sure you check several things. Did you review the rules of brainstorming, such as holding all evaluations until the brainstorm is done? Was there a recorder? Were all ideas written down? Did everyone participate? Remember, brainstorming doesn't only generate possible causes of a problem. It helps build people. The most serious thing that can go wrong is when someone in the group is put down. Did this happen in your group? Was it squared away?

Tables S-1, S-2, and S-3 are brainstorm lists for the fun projects. How many ideas did your group come up with? You can see piggybacking in Table S-1, Uses for old books: "Wall covering" is a piggyback on "wrap-

TABLE S-1
Brainstorm, uses for old books (Problem 2-1a).

USES FOR OLD BOOKS	
wrapping paper (arts and crafts)	rifle range backstop
wall covering	bird cage lining
fuel sources	novelty
place to hide key	booster chair
props (for construction)	gift
recycling	give to public library
garage sale	prop up leg on coffee table
container	landfill
give to Peace Corps	clothing
paper airplanes	cockroach food

TABLE S-2
Brainstorm, uses for discarded Christmas trees (Problem 2-1a).

USES FOR DISCARDED CHRISTMAS TREES

needles in a sachet pillow	use for smoke signals
needles in potpourri	let kids use for tree house
cones for decoration	boughs in camping mattress
roll cones in peanut butter for birds	cover for birds
wind break	protection for rabbits
insulation around foundation of house	on buoy in lake
tie food on for squirrels	use for hide and seek
use as scarecrow in garden	use for umbrella
grind up for mulch	cut up for firewood
cut off branches to cover flower beds	use several to build fence
start bonfire	use in erosion control

TABLE S-3
Brainstorm, uses for old tires (Problem 2-1a).

USES FOR OLD TIRES

recycle—raw material	obstacle course
tire swing	cut up and hook together for mats
sandals	bumpers on cars
planter	playground equipment
erosion control	doghouse
dock bumper	garden border
footholes (football team)	sled in winter
target—ball throwing	hula hoops
insulation	chair
tire races	shoes and boots
recap	buoys, floating apparatus
raft	flooring
bumpers on race course	tower building

TABLE S-4
Brainstorm, training program (Problem 2-1b).

GENERAL TRAINING ISSUES	
everyone is a sales representative	hospitalization
length of training	suggestion system
benefits	vacation scheduling
government regulations	drug testing
attendance	sick days
tardiness	employee involvement
leaving early	major customers
EPA regulations	major suppliers
handling returns	policy on customer complaints
retirement	OSHA
legal liability	customer surveys
discipline procedures	internal customers
promotion	appraisals

ping paper." In Table S-2, Uses for discarded Christmas trees, other examples of piggybacking are "cover for birds" and "protection for rabbits." Do you see other examples?

Problem 2-1b.

The brainstorm list for developing a training program will depend on what service industry, or even what department, your group is part of. When we did our brainstorm (see Table S-4), we purposely thought of general training issues. The ideas are not in any specific order, but you will notice evidence of piggybacking.

Problem 2-2.

Figure S-1 is a cause and effect diagram for disorganized inventory. How does it compare with the one you developed?

This diagram has five main bones: method, employee, material, environment, and supervision. You should include at least bones for method, employee, and material. You may have named differently the bones we

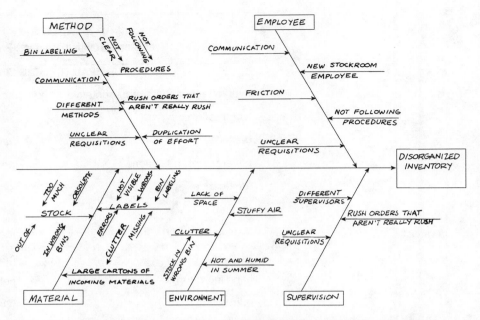

Figure S-1. Cause and effect diagram for disorganized inventory (Problem 2-2).

called supervision and environment. You may also have put the causes under different bones. Don't expect your C and E diagram to match exactly the one we drew.

Keep two things in mind as you study our C and E diagram. First, note that the bone for "supervision" has only three ideas. Somehow the brainstorm group didn't spend much thought on this area. The group should do another brainstorm—a second pass—to fill in this bone. However, if the group is new and still lacks confidence in the problem-solving process, they might want to put "supervision" on hold while they work on other aspects of the problem. Later, when the group has gained more confidence, they can go back and examine this area. This would be especially appropriate if the group works for the "supervision"! Eventually, though, they need to come back to "supervision" because an important part of the problem may lie in this area.

Second, the cause and effect diagram is a working diagram for a group to use as it tries to see how various ideas relate to each other. There's plenty of room for discussion on how exactly to organize the ideas. Some ideas will fit under more than one main bone. For example, we wrote "Unclear requisitions" under three main bones: "method," "employee," and "supervision." The cause and effect diagram is a continuation of the brainstorming process where we are trying to find possible causes of problems. So the rules of brainstorming apply here, too. If a member of the group wants to write a cause on a particular main bone, it should be written there even if the rest of the group disagrees. Remind the group to hold off evaluations for the end of the session.

"Training" is another main bone that the brainstorm group could have considered for this problem. Other problems may require other special bones in addition to the five major ones.

Problem 2-3.

Pareto analysis is an excellent tool to help your team determine where lie the best opportunities for reducing data entry errors. Data from the report show what types of data entry errors there are and how often they are occurring.

First, rank these errors by largest frequency on the top part of the Pareto diagram worksheet. Next, determine the cumulative frequencies and record them on the worksheet. You can combine "Reversed digits" and "Wrong customer address" into one category, "Others," because these two items together are less than 10% of the total of all frequencies. See Figure S-2.

Now you can draw the Pareteo diagram on the graph portion of the worksheet. You will have eight equal-width bars.

Determine the cumulative percentages. Refer to Figure S-2. Draw them on the diagram. The completed diagram should look like Figure S-3.

In reviewing this Pareto diagram, it is obvious that "Incorrect catalog number" and "Price change" are the largest problem areas. You can strongly recommend to your manager that concentrating improvement efforts on these two items should reduce a large percentage of data entry errors.

MODULE 3: QUALITY IMPROVEMENT TOOLS

Problem 3-1.

Figure S-4 is a process flow chart for the morning's activities. How does it compare to the one you drew?

Step 1 is a type of storage because the printout is already on the desk waiting for the clerk. Step 6 is drawn as an inspection step because the

Figure S-2. Top part of worksheet for Pareto diagram of data entry errors (Problem 2-3).

PARETO DIAGRAM – WORKSHEET

Category	Frequency	Cumulative frequency	Cumulative percentage
Incorrect catalog number	58	58	23
Price change	47	105	42
Misspelling	43	148	59
Unclear handwriting	31	179	71
Wrong quantity	24	203	81
Wrong price	17	220	88
Incorrect description	13	233	93
Other	18	251	100

Figure S-3. Completed Pareto diagram of data entry errors (Problem 2-3).

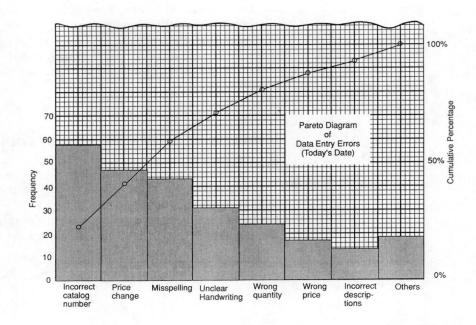

Figure S-4. Flow chart of morning's activities in billing office (Problem 3-1).

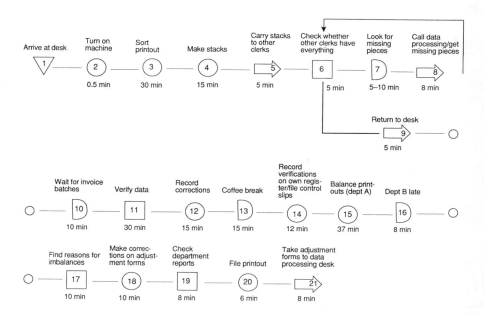

clerk is checking whether the next three people have all the necessary information they need to do their operations. If they do, they can proceed with their operations and you can return to your work. Step 7 is the type of delay that is both irritating and a waste of time and effort. It would be a good place for improvement. Step 8 is a type of move because you have to go to data processing to get any missing data inputs. You could also consider this a delay from the standpoint of your own work. Any step that involves verification should be drawn as an inspection step because we are looking to see whether there are errors or things left out.

What steps would you target for improvement?

Problem 3-2.

The task team has agreed that careful, initial planning will most likely ensure success of the project. Ideas they suggested under "purpose" included "to become more active in the community"; "to help people who need help"; and "to put love into action."

Their brainstorm for major headings produced the following headings. What headings did you use?

1. What will our people actually be doing?
2. Training.
3. Logistics.
4. Building usage.

5. Legal concerns/codes.
6. Spiritual matters.
7. Direct assistance.

Figure S-5 is a storyboard that a planning session developed. How does it compare to the one you did?

At one point during the storyboard session, a team member suggested that the card "supplies for program" under heading 7 should be a heading by itself because it would make listing of supplies easier. The team agreed, and added an eighth heading, "supplies for program."

The next step for this team is to evaluate their storyboard. Once the evaluation phase is complete, the team can start involving the rest of the congregation in further planning for this project.

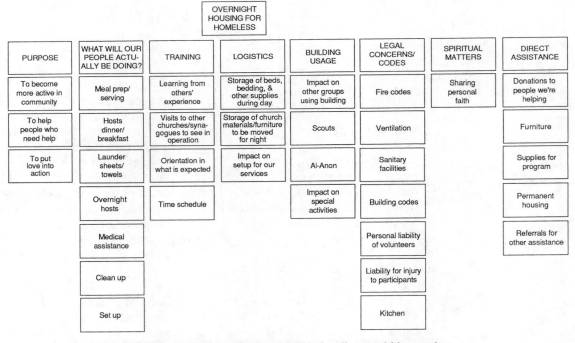

Figure S-5. Storyboard showing purpose and other possible headings and idea cards posted under appropriate headings (Problem 3-2).

Problem 3-3.

Figure S-6 is the scatter diagram of the data. The diagram shows the plotted variables and the two median lines.

Figure S-6. Scatter diagram of number of gallons used versus number of miles driven (Problem 3-3).

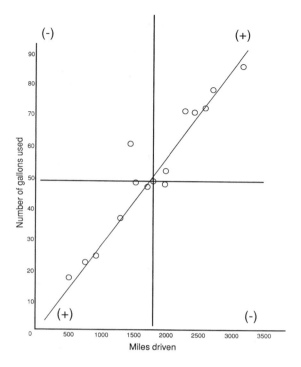

The results of our counts are

Quarter	Count
Count from right	+6
Count from top	+5
Count from left	+4
Count from bottom	+5
Total	+20

As we move a ruler vertically in from the right side of the diagram, we meet six points in the upper quarter to the right of the vertical median line. The very next point we meet is still to the right of the vertical median but is *below* the *horizontal* median line and in a different quarter. Therefore, we stop counting at 6. The count for this corner is +6

When we move the ruler down from the top toward the horizontal median line, we meet five points in the upper right quarter. The sixth point is also above the horizontal line but to the left of the vertical median line. So, we do not count it. Our total is +5. You will notice that we have already counted these same five points when we moved the ruler in from the right. This is not unusual when doing the corner count test.

Counting from the left, we have four points. The fifth point lies above the horizontal median line, and we don't count it. Coming up from the bottom, we find five points before meeting the sixth point, which lies to the right of the vertical median line.

The total, after dropping the plus or minus sign, is 20. This is higher than the statistician's value of "11." Therefore, we can conclude that there really is a straight-line relationship.

You probably already expected a straight-line relationship in this problem. The automobiles are all the same model and brand. We know from experience that the farther we drive, the more gas we must put into the tank. We also expect that if we double the mileage driven, we will have to put in about twice as many gallons of fuel: This means the relationship should be straight-line.

We drew a straight line through the data on the diagram by using the "eyeball" method (we drew it where it looked about right). In comparing the data to this line, one point looks odd. Point number 9, representing 1,453 miles and 61.2 gallons, looks different from the others. This point could be different because of an error in recording the data. But, if point number 9 is not an error, apparently something is wrong with that car, because it is using more fuel than the others. If you calculate miles per gallon (mpg) as shown in Table S-5, you can see that this car has the lowest mpg. This vehicle is a good candidate for a maintenance check.

TABLE S-5
Calculations of miles per gallon (mpg)

NUMBER OF MILES DRIVEN	NUMBER OF GALLONS USED	MPG	NUMBER OF MILES DRIVEN	NUMBER OF GALLONS USED	MPG
1523	48.6	31.3	1453	61.2	23.7
1723	46.8	36.8	1980	47.1	42.0
2723	78.7	34.6	2452	70.8	34.6
2602	72.7	35.8	927	25.1	36.9
1805	49.2	36.7	2012	52.7	38.2
510	18.2	28.0	2303	71.5	32.2
762	23.1	33.0	1302	37.1	35.1
3212	86.2	37.3			

MODULE 4: FREQUENCY HISTOGRAMS

Problem 4-1.

In this problem, you are comparing four manual processes that result in the same service, a double-dip ice cream cone. Any time you have a manual operation where you are weighing or measuring something, you will have variation. Comparisons of the histograms will be easier to make if you use the same scale for the frequency and the same number and size of intervals for all the histograms. Based on the data given in the problem, your histograms should look something like those in Figures S-7 through S-11.

By lining up the histograms in a vertical column in front of you, it will be easy to compare them.

(a) There are large differences between the averages of the first four histograms. Figure S-7 (A, B, C, D) has an average of about 5.5; Figure S-8 (E, F, G, H) has 6.1; Figure S-9 (I, J, K, L) has 6.9; and Figure S-10 (M, N, O, P) has 4.8.

(b) The spreads of Figures S-7, S-8, and S-10 are about the same, even though the measurements for the weights are clustered around different averages. The spread of Figure S-9 is much less than the other three.

(c) The fifth histogram (Figure S-11, columns A, E, I, M) is a composite of weights from each of the first four. In developing this histogram, we broke one of the rules for constructing histograms: use data from only one source. We did this to make a point. The point is that although histograms are easy to use, they cannot always detect when something is wrong in a process. The histogram in Figure S-11 gives no indication that you are dealing with data from more than one source. Notice that the largest reading is the very largest reading of the other four; the smallest reading is almost as small as the very smallest. Therefore, the spread of this histogram is much wider than the others.

Problem 4-2.

(a) The frequency histogram, including all the worksheet calculations, is shown in Figure S-12.

(b) The average from eyeballing the frequency histogram is about 3 minutes and 30 seconds. If you chose a value a little more or less than this, that's O.K.

(c) Our frequency histogram does not appear to be "bell-shaped." The highest point of the graph is at 2 minutes, 30 seconds. There is a second peak at 6 minutes and 30 seconds. A bell-shaped graph should have only one peak, and it should be in the center, not pushed to the left side as ours is. Does yours look about the same? Remember that if you chose a different number of intervals than we did, or if you have bound-

Figure S-7. Frequency histogram of ice cream cone weights, columns A, B, C, and D (Problem 4-1).

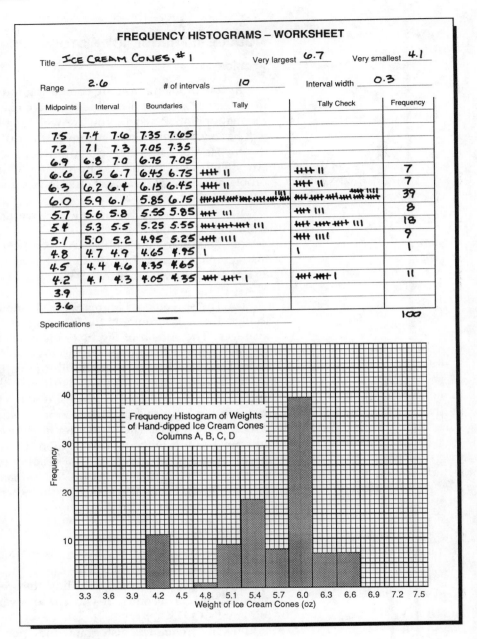

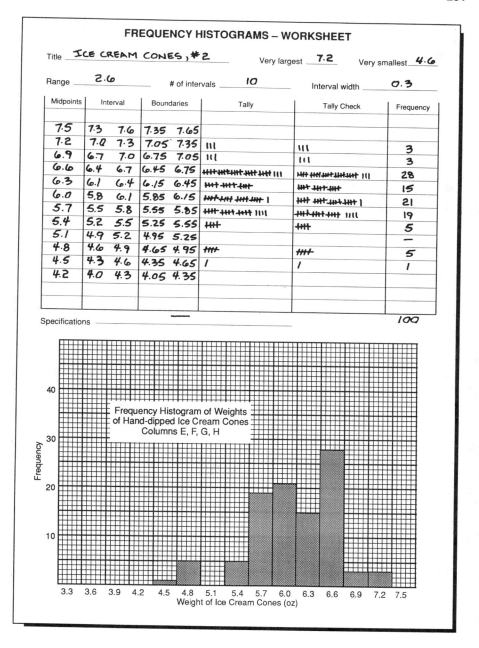

Figure S-8. Frequency histogram of ice cream cone weights, columns E, F, G, and H (Problem 4-1).

Figure S-9. Frequency histogram of ice cream cone weights, columns I, J, K, and L (Problem 4-1).

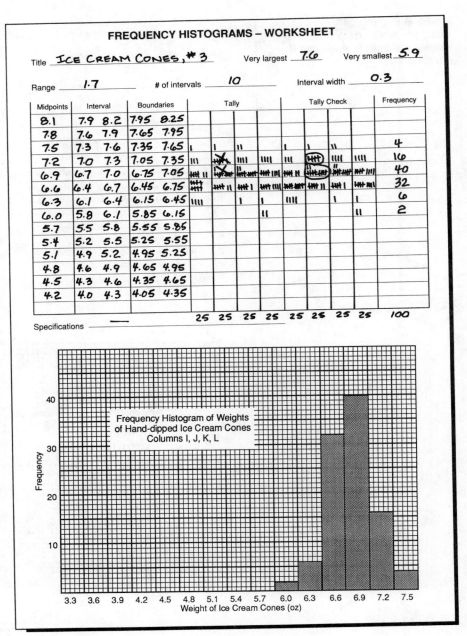

Figure S-10. Frequency histogram of ice cream cone weights, columns M, N, O, and P (Problem 4-1).

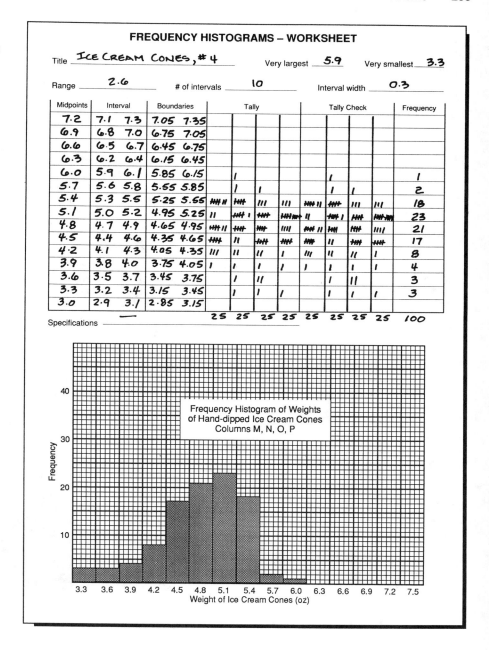

Figure S-11. Frequency histogram of ice cream cone weights, columns A, E, I, and M (Problem 4-1).

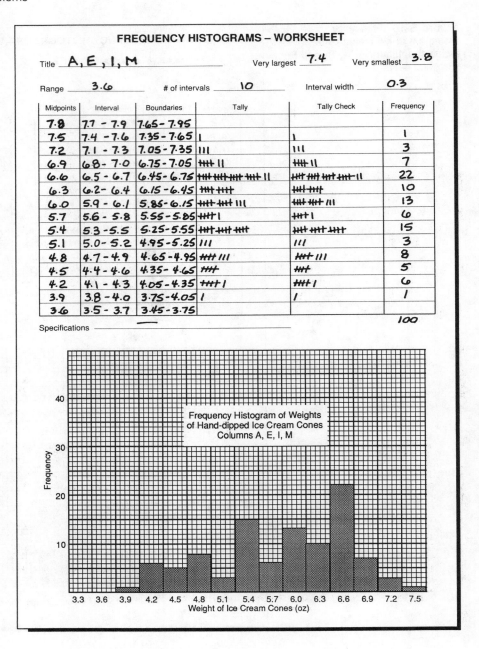

Figure S-12. Frequency histogram of transaction times for drive-through window (Problem 4-2).

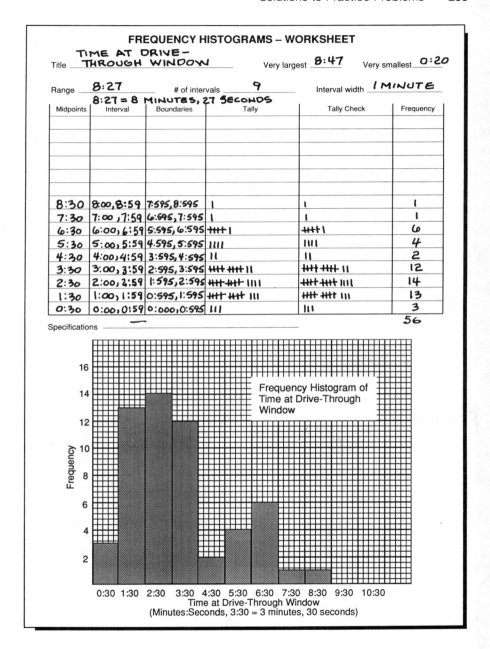

aries and midpoints different from ours, your graph will look a little different.

When we think about the kind of data we have, we really would not expect it to be bell-shaped. Most of the times should be short (for the sake of the customer!). A few customers will have lengthy transactions. Perhaps they send the pneumatic tube back and forth several times before finishing. So a few long times are not unexpected. However, since there cannot be any times less than zero, the graph will not be able to spread out to the left as a "bell" would do.

(d) There are three suspicious points—those in the zero minutes, 30 seconds interval. The customer cannot send the pneumatic tube in, the teller process the transaction, and return it in less than 1 minute. These three observations probably represent customers who asked a question and then left without any financial transaction. Or they simply left the window.

Our graph is curious in that it has two major peaks, one at 2 minutes, 30 seconds and another at 6 minutes, 30 seconds. This could be from sampling variation. There are only 56 observations. However, since it aroused our suspicion, we should collect more data and look to see if the same thing happens again.

Problem 4-3.

(a) A checklist checksheet for buying a personal computer for home business use might look like the following.

Checklist for Purchasing PC

12 MHz speed minimum	____
20 MB hard drive minimum.............	____
VGA color monitor......................	____
IBM compatible	____
3.5-inch floppy drive	____
Surge protect	____
24-pin dot-matrix printer	____
Maximum price of $4000.00	____
Brown woodwork station on wheels	____
Modem	____

(b) The following is a checksheet for possible defects or problems with a PC.

Defect/Problem Checksheet for PC

Monitor
- Dim .... _____
- Fuzzy .... _____
- Electrical snaps .... _____
- Doesn't work at all .... _____
- Other .... _____

Keyboard
- Keys don't function .... _____
- Works intermittently .... _____
- Types in wrong language .... _____
- Other .... _____

Central unit
- Doesn't start .... _____
- Shuts down unexpectedly .... _____
- Loses data on hard disk .... _____
- Other .... _____

Floppy drive
- Doesn't engage .... _____
- Doesn't copy in .... _____
- Doesn't copy out .... _____
- Other .... _____

Printer
- Doesn't turn on .... _____
- Ejects extra page .... _____
- Page number in wrong location .... _____
- Some characters incorrect .... _____
- Other .... _____

(c) A location checksheet for use of file cabinets might look like Figure S-13.

Figure S-13. Location checksheet for use of file cabinets (Problem 4-3).

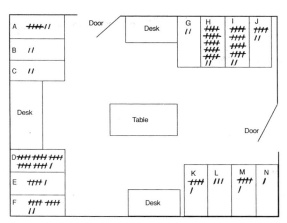

(d) The intervals you use on your frequency histogram worksheet for miles per gallon will depend on the size car you drive. For an economy car you could run the scale from 30 to 50 mpg. But a gas-guzzler may require that the horizontal scale run from 10 to 20 mpg. Our solution is for a car that gets between 25 to 35 mpg. See Figure S-14.

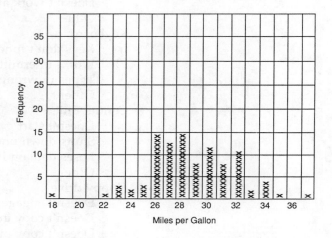

Figure S-14. Frequency histogram checksheet for midsize automobile (Problem 4-3).

(e) The checksheet of Figure S-15 shows some of our favorite restaurants. How does the setup of your checksheet compare?

Figure S-15. Matrix checksheet showing my favorite restaurants and what's important to me as a customer (Problem 4-3).

	Anticoli's	Dragon Inn	Jay's	Mr. Tiffany's	Olive Garden
Service					
· Prompt	_____	_____	_____	_____	_____
· Courteous	_____	_____	_____	_____	_____
· Friendly	_____	_____	_____	_____	_____
· Unhurried	_____	_____	_____	_____	_____
Menu					
· Variety	_____	_____	_____	_____	_____
· Steaks	_____	_____	_____	_____	_____
· Seafood	_____	_____	_____	_____	_____
Convenient location	_____	_____	_____	_____	_____
Price					
· Low	_____	_____	_____	_____	_____
· Medium	_____	_____	_____	_____	_____
· High	_____	_____	_____	_____	_____
Lunch specials	_____	_____	_____	_____	_____
Atmosphere					
· Relaxed	_____	_____	_____	_____	_____
· Private	_____	_____	_____	_____	_____

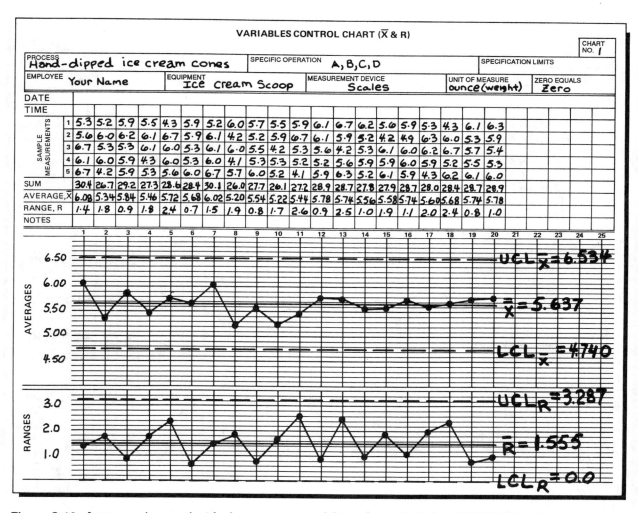

Figure S-16. Average and range chart for ice cream cone weights, columns A, B, C and D (Problem 5-1).

$\bar{\bar{R}} = \dfrac{31.1}{20} = 1.555$

$\bar{\bar{X}} = \dfrac{112.74}{20} = 5.637$

$UCL_R = D_4 \times \bar{R}$
$= 2.114 \times 1.555$
$= 3.287$

$UCL_{\bar{X}} = \bar{\bar{X}} + (A_2 \times \bar{R})$
$= 5.637 + (0.577 \times 1.555)$
$= 5.637 + 0.897$
$= 6.534$

$LCL_{\bar{X}} = 5.637 - 0.897$
$= 4.740$

MODULE 5: VARIABLES CONTROL CHARTS

Problem 5-1.

In Problem 4-1, you analyzed data on the weights of double-dip ice cream cones using frequency histograms. For this problem you are analyzing the same data with a different statistical technique, the average and range chart. Your control charts based on the data in Problem 4-1 should look like the ones in Figures S-16 through S-20.

(a) Just as for the frequency histograms, there are large differences between the overall averages ($\bar{\bar{X}}$'s) of the first four control charts. For Fig-

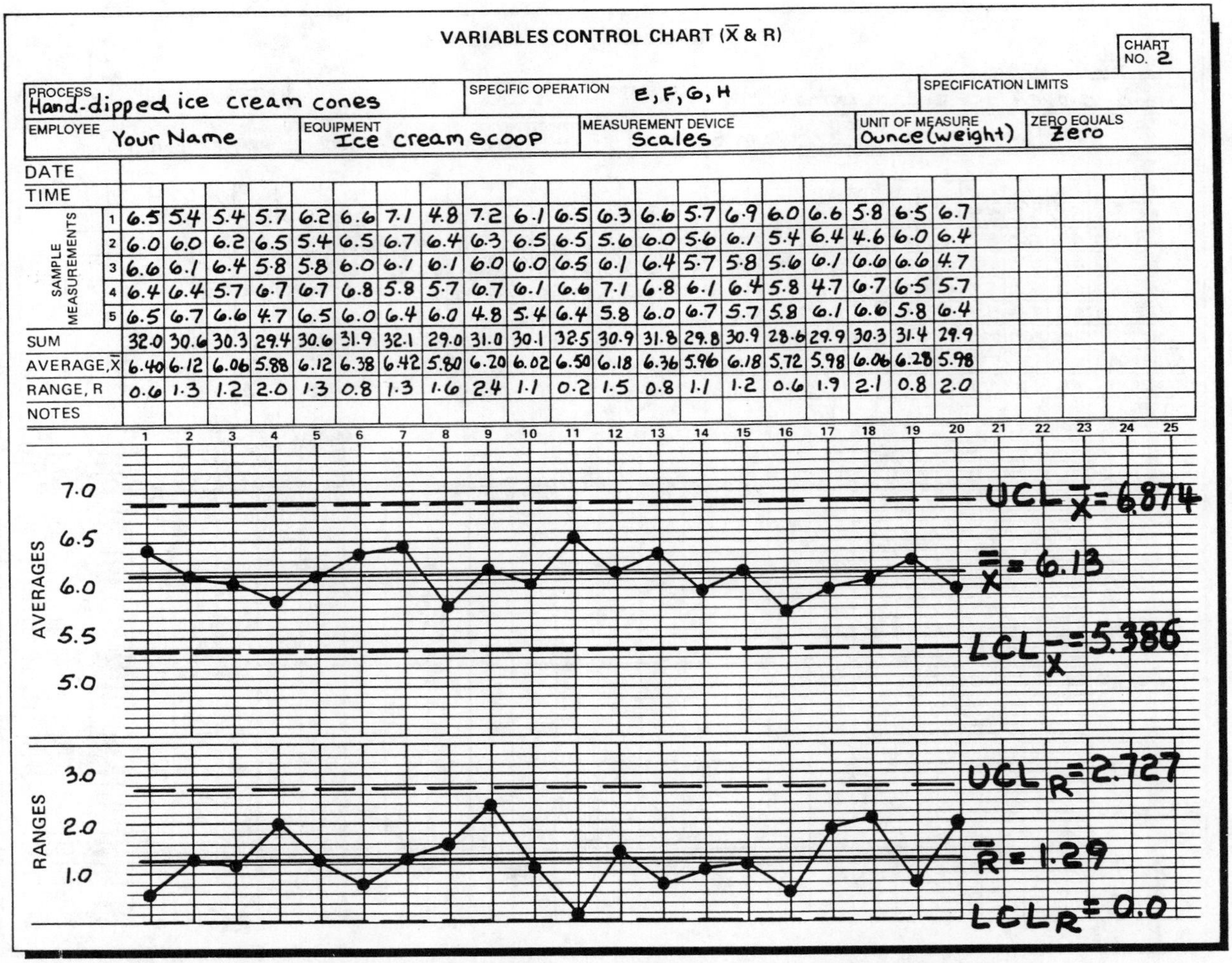

	1	2	3	4	5	6	7	8	9	10	11	12	13	14	15	16	17	18	19	20
Sample measurements 1	6.5	5.4	5.4	5.7	6.2	6.6	7.1	4.8	7.2	6.1	6.5	6.3	6.6	5.7	6.9	6.0	6.6	5.8	6.5	6.7
2	6.0	6.0	6.2	6.5	5.4	6.5	6.7	6.4	6.3	6.5	6.5	5.6	6.0	5.6	6.1	5.4	6.4	4.6	6.0	6.4
3	6.6	6.1	6.4	5.8	5.8	6.0	6.1	6.1	6.0	6.0	6.5	6.1	6.4	5.7	5.8	5.6	6.1	6.6	6.6	4.7
4	6.4	6.4	5.7	6.7	6.7	6.8	5.8	5.7	6.7	6.1	6.6	7.1	6.8	6.1	6.4	5.8	4.7	6.7	6.5	5.7
5	6.5	6.7	6.6	4.7	6.5	6.0	6.4	6.0	4.8	5.4	6.4	5.8	6.0	6.7	5.7	5.8	6.1	6.6	5.8	6.4
SUM	32.0	30.6	30.3	29.4	30.6	31.9	32.1	29.0	31.0	30.1	32.5	30.9	31.8	29.8	30.9	28.6	29.9	30.3	31.4	29.9
AVERAGE, $\overline{X}$	6.40	6.12	6.06	5.88	6.12	6.38	6.42	5.80	6.20	6.02	6.50	6.18	6.36	5.96	6.18	5.72	5.98	6.06	6.28	5.98
RANGE, R	0.6	1.3	1.2	2.0	1.3	0.8	1.3	1.6	2.4	1.1	0.2	1.5	0.8	1.1	1.2	0.6	1.9	2.1	0.8	2.0

Figure S-17. Average and range chart for ice cream cone weights, columns E, F, G, and H (Problem 5-1).

$$\overline{R} = \frac{25.8}{20} = 1.29$$

$$\overline{\overline{X}} = \frac{122.60}{20} = 6.13$$

$$UCL_R = D_4 \times \overline{R} = 2.114 \times 1.29 = 2.727$$

$$UCL_{\overline{X}} = \overline{\overline{X}} + (A_2 \times \overline{R}) = 6.13 + (0.577 \times 1.29) = 6.13 + 0.744 = 6.874$$

$$LCL_{\overline{X}} = 6.13 - 0.744 = 5.386$$

ure S-16 (A, B, C, D), the $\overline{\overline{X}}$ is 5.637; for Figure S-17 (E, F, G, H), it is 6.13; for Figure S-18 (I, J, K, L), it is 6.829; and for Figure S-19 (M, N, O, P), it is 4.828. So the answer should be yes. The average for Figure S-20 (A, E, I, M) is 5.874. It lies between the two low averages for Figures S-16 and S-19 and the two high ones of Figures S-17 and S-18. This is to be expected since the data for Figure S-20 is a composite of columns from each of the other four charts.

(b) The overall averages for the control charts follow the same pattern that you saw in the averages for frequency histograms. To find the averages for the frequency histograms, we eyeballed the histograms. To find

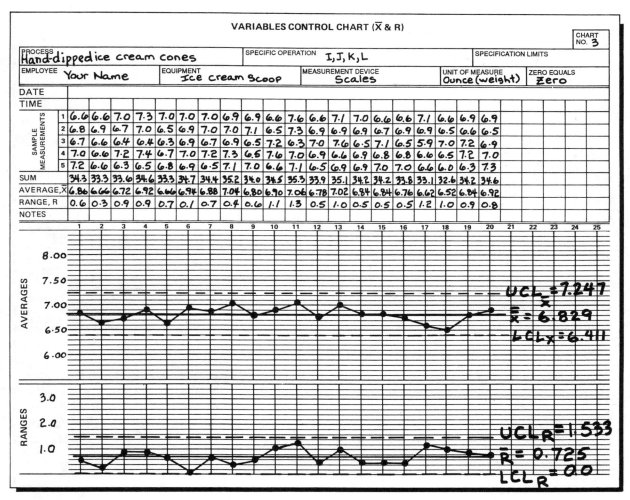

Figure S-18. Average and range chart for ice cream cone weights, columns I, J, K, and L (Problem 5-1).

$$\bar{R} = \frac{14.5}{20} = 0.725$$

$$\bar{\bar{X}} = \frac{136.58}{20} = 6.829$$

$$UCL_R = D_4 \times \bar{R}$$
$$= 2.114 \times 0.725$$
$$= 1.533$$

$$UCL_{\bar{X}} = \bar{\bar{X}} + (A_2 \times \bar{R})$$
$$= 6.829 + (0.577 \times 0.725)$$
$$= 6.829 + 0.418$$
$$= 7.247$$

$$LCL_{\bar{X}} = 6.829 - 0.418$$
$$= 6.411$$

the $\bar{\bar{X}}$'s on the control charts, we calculated the average of the subgroup averages. Also notice that the control chart averages are carried to three decimal places.

We obtained the averages for the frequency histograms and the control charts from the same data, but the statistical technique used for the control chart gives a more precise answer than the eyeball method used for the histograms. Both methods give *estimates* of the true process average, but the control chart estimate is *more* exact.

(c) The control chart for A, E, I, M (Figure S-20) shows points out of control on the averages chart. This is a clear indication that assignable

Solutions to Practice Problems

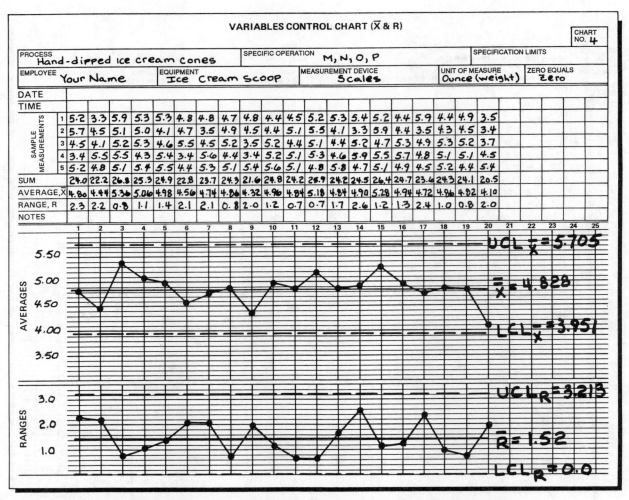

Figure S-19. Average and range chart for ice cream cone weights, columns M, N, O, and P (Problem 5-1).

$$\overline{R} = \frac{30.4}{20} = 1.52$$

$$\overline{\overline{X}} = \frac{96.56}{20} = 4.828$$

$$UCL_R = D_4 \times \overline{R}$$
$$= 2.114 \times 1.52$$
$$= 3.213$$

$$UCL_{\overline{X}} = \overline{\overline{X}} + (A_2 \times \overline{R})$$
$$= 4.828 + (0.577 \times 1.52)$$
$$= 4.828 + 0.877$$
$$= 5.705$$

$$LCL_{\overline{X}} = 4.828 - 0.877$$
$$= 3.951$$

causes are present and that the process is out of control. This should indicate to you that the average and range chart is far better than the frequency histogram in detecting when something is wrong with the process and/or with the data used to develop the control chart.

The histogram in Figure S-19 gives no indication that there is anything unusual about the data. In contrast, the average and range chart in Figure S-20 shows very clearly that something is wrong. You already know that the data for this chart are a composite of four scooping operations. But even if you didn't have information, the control chart would alert you to the presence of assignable causes, whereas the histogram probably would not.

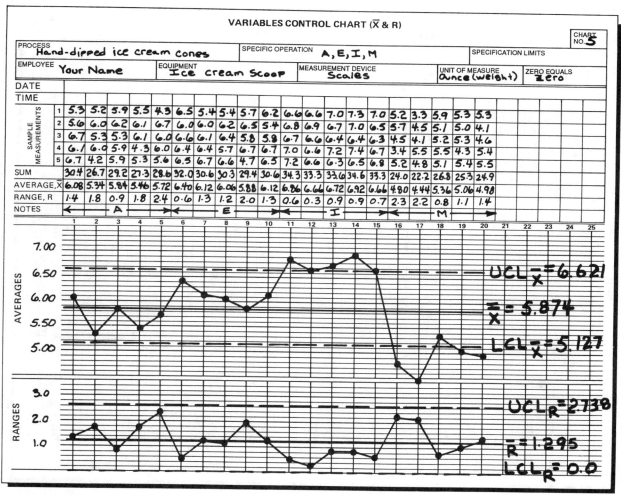

Figure S-20. Average and range chart for ice cream cone weights, columns A, E, I, and M (Problem 5-1).

$\bar{R} = \dfrac{25.9}{20} = 1.295$

$\bar{\bar{X}} = \dfrac{117.48}{20} = 5.874$

$UCL_R = D_4 \times \bar{R}$
$= 2.114 \times 1.295$
$= 2.738$

$UCL_{\bar{X}} = \bar{\bar{X}} + (A_2 \times \bar{R})$
$= 5.874 + (0.577 \times 1.295)$
$= 5.874 + 0.747$
$= 6.621$

$LCL_{\bar{X}} = 5.874 - 0.747$
$= 5.127$

Problem 5-2.

Figure S-21 is the median and range chart we constructed from the decibel readings in Problem 5-2. How does your chart compare?

In developing this chart, we grouped the readings into sets of 5. This way, we have 20 medians that we plotted on the chart. Since there are 20 medians, the median of the medians ($\tilde{\tilde{X}}$) is the value of the median between the tenth and the eleventh subgroup medians (count up from the smallest median value). Figure S-21 shows that the value of the tenth subgroup median is 28 and the value of the eleventh median is 29. So the $\tilde{\tilde{X}}$ is 28.5, halfway between these two values, $(28 + 29) \div 2 = 28.5$.

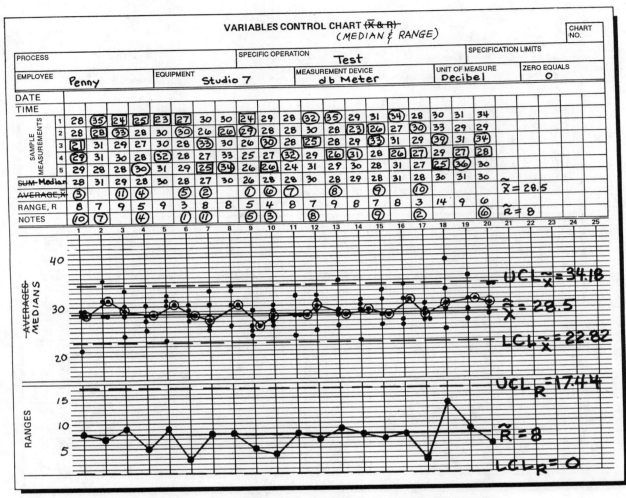

Figure S-21. Median and range chart for test measurements of radio equipment (Problem 5-2).

$\widetilde{R}$ #10=8, #11=8,
so $\widetilde{R}$=8

$\widetilde{\overline{X}}$ #10=28, #11=29,
so $\widetilde{\overline{X}} = \frac{(28+29)}{2} = 28.5$

$UCL_R = \widetilde{D}_4 \times \widetilde{R}$
$= 2.18 \times 8$
$= 17.44$

$UCL_{\widetilde{X}} = \widetilde{\overline{X}} + (\widetilde{A}_2 \times \widetilde{R})$
$= 28.5 + (0.71 \times 8)$
$= 28.5 + 5.68$
$= 34.18$

$LCL_{\widetilde{X}} = 28.5 - 5.68$
$= 22.82$

The tenth range value is 8 and the eleventh range value is also 8, so the median of the ranges ($\widetilde{R}$) is 8.

Remember that the $\widetilde{D}_4$ and the $\widetilde{A}_2$ factors you use to calculate the control limits for ranges and medians are not the same as for calculating control limits for averages and ranges. Also keep in mind that the values of the $\widetilde{D}_4$ and the $\widetilde{A}_2$ factors are for a sample size of 5.

(a) In Figure S-21, we see that all the medians and ranges lie within the control limits. Therefore, we can say that this process is in control.

(b) When we look at the median portion of the chart, we see that several individual decibel readings fall outside both the upper and lower control limits for medians. This does *not* mean the process is out of statis-

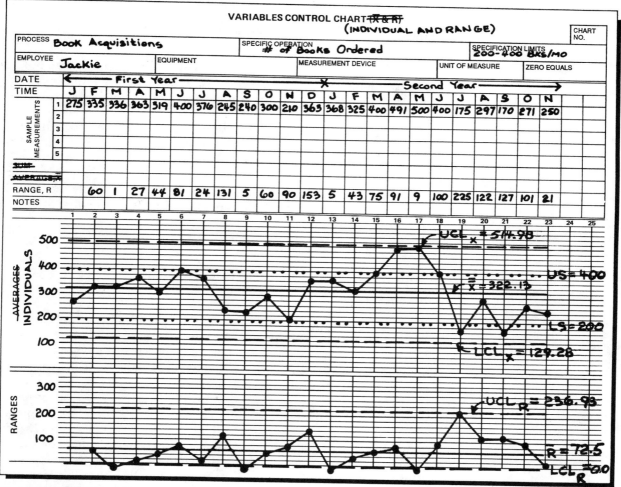

Figure S-22. Individual and range chart for number of books ordered. Individuals chart has control limits and specifications drawn in (Problem 5-3).

$$\bar{R} = \frac{1595}{22} = 72.5$$

$$\bar{\bar{X}} = \frac{7409}{23} = 322.13$$

$$UCL_R = D_4 \times \bar{R}$$
$$= 3.268 \times 72.5$$
$$= 236.93$$

$$UCL_X = \bar{\bar{X}} + (A_2 \times \bar{R})$$
$$= 322.13 + (2.66 \times 72.5)$$
$$= 322.13 + 192.85$$
$$= 514.98$$

$$LCL_X = 322.13 - 192.85$$
$$= 129.28$$

tical control. No corrective action is required in such a situation because no *medians* are outside the control limits.

Problem 5-3.

The individual and range (X-R) chart you constructed from the book acquisitions data should look like the chart in Figure S-22. Remember, the value of the D_4 factor you use to calculate the upper control limit for ranges is the same as the one for the average and range chart when the sample size is 2. (See Figure 5-12; n=2.) The A_2 factor is 2.66.

When you constructed this chart, you had 23 individual points representing orders for 23 months. In this problem there are only 22 ranges because there is always one less range than there are individual readings in the individual and range chart. Because this is an X-R chart, you *may* draw in the specifications as we did.

(a) The individual and range chart is appropriate in this case because the acquisitions department has only one reading on the number of books ordered each month. Also there is an interval of time, one month, between readings.

(b) The book ordering process is not consistently meeting the specifications of 200–400 books ordered per month. In April and May of the second year, more than 400 books were ordered; in July and September of that year, fewer books were ordered.

(c) Even though the ordering process is not consistently meeting the specifications, it is in control. No ranges or individual months' orders fall outside the control limits.

MODULE 6: ATTRIBUTES CONTROL CHARTS

Problem 6-1.

The inspection data in this problem are the number of *defectives*, not defects, in samples of 48 panels. You can use these data to construct a percent defective chart. Your chart should look like the one in Figure S-23. Compare your calculations for the average percent defective ($\bar{p}$) and the control limits with those we give.

At issue is whether assignable causes or system causes are responsible for the dirt and grease spots on the panels. If there are assignable causes, somewhere in the storage/loading area, something is probably going wrong. However, if there are system causes, management of the improvement center needs to make improvements. Because the average percent defective is high (52.1%), most people would blame the employees in the loading area. But the control chart shows that all the percents defective (p's) are inside the control limits. Statistically, we can say that this high percentage of dirt and grease spots on panels is somehow built into the system. There cannot be significant and lasting improvement without changes in the system. This requires action and guidance from management.

Problem 6-2.

For this problem, you must construct three c-charts. You are counting the number of errors in the inspection unit, which is the 2-day period when the toll agency checks transactions. When you first set up these c-charts, they should look like Figures S-24, S-25, and S-26. In constructing

Figure S-23. p-chart of dirt and grease spots on panels at loading/storage (Problem 6-1).

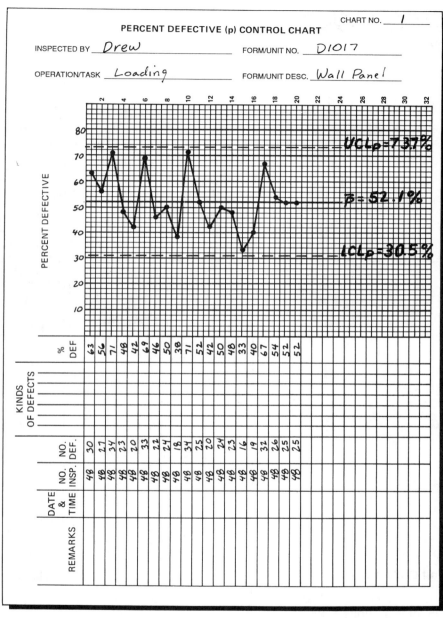

$$n = 48$$
$$\text{Total defective} = 500$$
$$\text{Total inspected} = 960$$
$$\bar{p} = \frac{500}{960} \times 100\% = 52.1\%$$
$$UCL_p = 52.1 + 3\sqrt{\frac{52.1\,(100-52.1)}{48}}$$
$$= 52.1 + 21.6$$
$$= 73.7\%$$
$$LCL_p = 52.1 - 21.6$$
$$= 30.5\%$$

Figure S-24. c-chart for toll collector error rates, A (Problem 6-2).

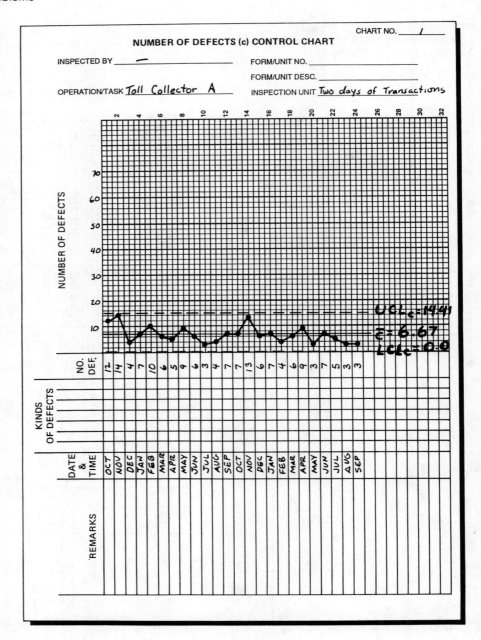

Toll Collector A

Total c's = 160
No. of insp. units = 24

$$\bar{c} = \frac{160}{24} = 6.67$$

$$\begin{aligned}
UCL_c &= \bar{c} + 3\sqrt{\bar{c}} \\
&= 6.67 + 3\sqrt{6.67} \\
&= 6.67 + 7.74 \\
&= 14.41
\end{aligned}$$

$$LCL_c = 0$$

Process is stable.

Figure S-25. c-chart for toll collector error rates, B (Problem 6-2).

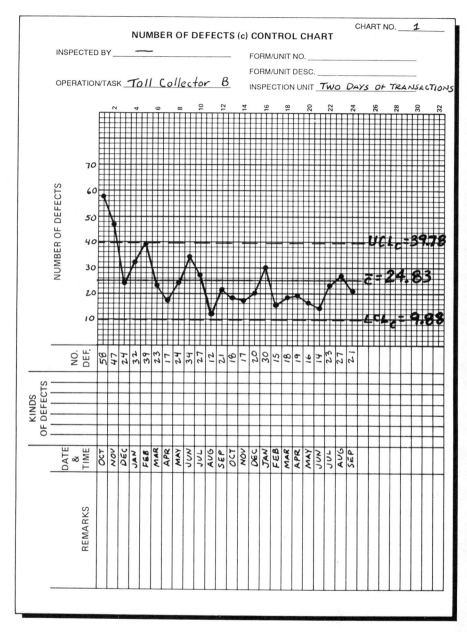

Toll Collector B

Total c's = 596
No. of insp. units = 24

$\bar{c} = \dfrac{596}{24} = 24.83$

$UCL_c = \bar{c} + 3\sqrt{\bar{c}}$
$= 24.83 + 3\sqrt{24.83}$
$= 24.83 + 14.95$
$= 39.78$

$LCL_c = 24.83 - 14.95$
$= 9.88$

If following rules exactly, delete #1 and #2.

new $\bar{c} = \dfrac{491}{22} = 22.32$

$UCL_c = 36.49$
$LCL_c = 8.15$

One point, #5, out of control. Process is not stable.

If deleting first 5 months, as learning period:

new $\bar{c} = \dfrac{396}{19} = 20.84$

$UCL_c = 34.54$
$LCL_c = 7.14$

From month 6 on, process is stable.

Figure S-26. c-chart for toll collector error rates, C (Problem 6-2).

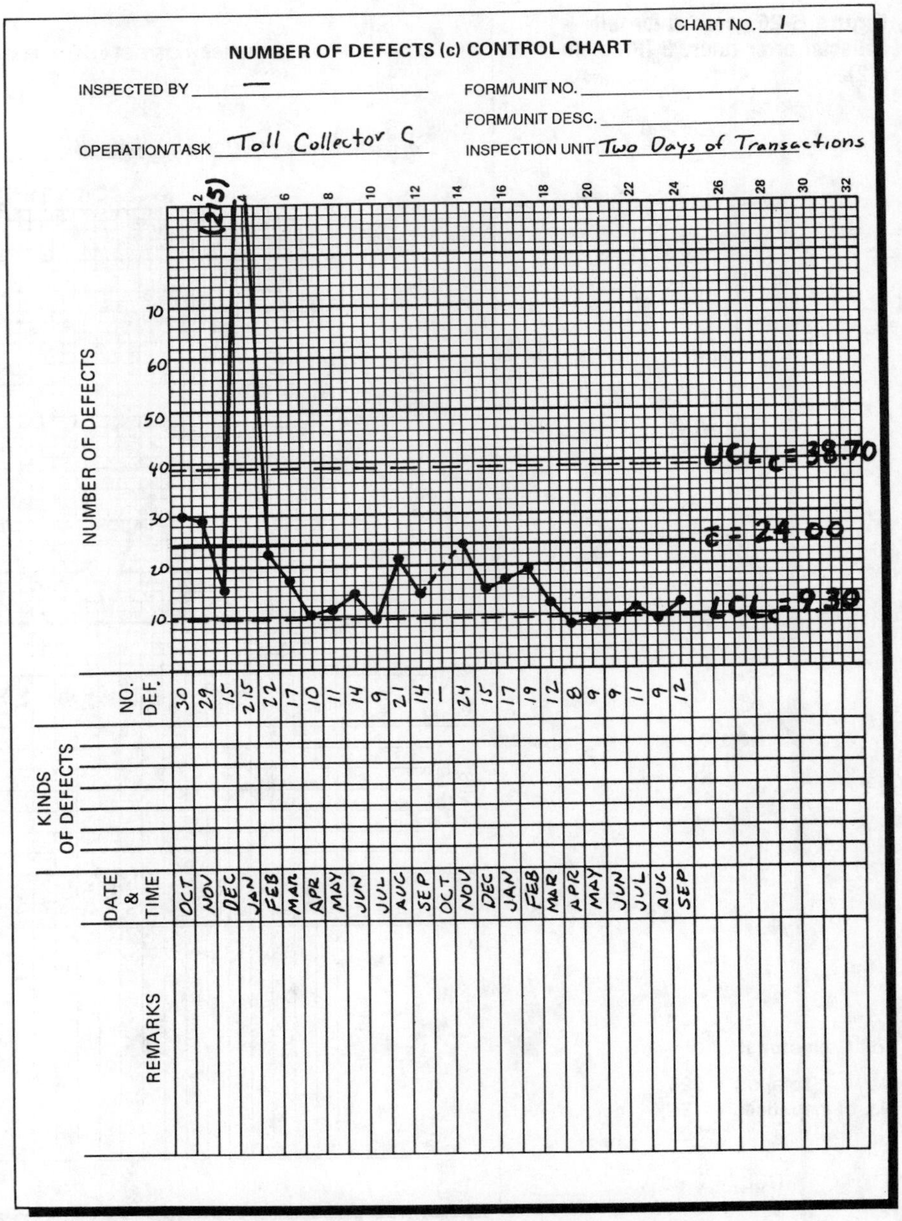
CHART NO. 1
NUMBER OF DEFECTS (c) CONTROL CHART
INSPECTED BY —
FORM/UNIT NO.
FORM/UNIT DESC.
OPERATION/TASK Toll Collector C
INSPECTION UNIT Two Days of Transactions
NUMBER OF DEFECTS
(215)
UCLc = 38.70
c̄ = 24.00
LCLc = 9.30
NO. DEF.
30 29 15 215 22 17 10 11 14 9 21 14 — 24 15 17 19 12 8 9 9 11 9 12
KINDS OF DEFECTS
DATE & TIME
OCT NOV DEC JAN FEB MAR APR MAY JUN JUL AUG SEP OCT NOV DEC JAN FEB MAR APR MAY JUN JUL AUG SEP
REMARKS

Toll Collector C

$$\text{Total c's} = 552$$
$$\text{No. of insp. units} = 23$$
$$\bar{c} = \frac{552}{23} = 24.00$$
$$UCL_c = \bar{c} + 3\sqrt{\bar{c}}$$
$$= 24.0 + 3\sqrt{24.0}$$
$$= 24.0 + 14.70$$
$$= 38.7$$
$$LCL_c = 24.0 - 14.70$$
$$= 9.30$$

Sample #4 is way out. Delete *only* sample #4.

$$\text{new } \bar{c} = \frac{337}{22} = 15.32$$
$$UCL_c = 15.32 + 3\sqrt{15.32}$$
$$= 27.06$$
$$LCL_c = 15.32 - 11.74$$
$$= 3.58$$

Two points, #1 and #2, out of control. Process is not stable.
If deleting first 5 months, as learning period:

$$\text{new } \bar{c} = \frac{241}{18} = 13.39$$
$$UCL_c = 24.37$$
$$LCL_c = 2.41$$

From month 6 on, process is stable.

these c-charts, use the same scale for the three charts. In that way it will be easier to see how the error rates compare.

(a) When first establishing the control charts for toll collectors B and C, you find there are points outside the control limits. According to the rules for setting up c-charts, throw out those one or two c's and the samples from which they come. Completely recalculate the $\bar{c}$ (average number of errors for the process). Using the new $\bar{c}$, calculate new control limits.

In the case of collector B, throw out the readings for the first 2 months. Then the new $\bar{c}$ for B is 22.32; the new UCL_c is 36.49; and the new LCL_c is 8.15. However, there is still one point outside the upper control limit.

In the case of collector C, there are six points outside the control limits: one "wild" point of 215 above the upper control limit in month 4 and five points outside the lower control limit for months 10, 19, 20, 21, and 23. When we made the recalculations for this chart, we broke one of the rules for establishing c-charts because we removed only the one wild reading of 215. But we did this to find whether the new c-chart would be in control if this one very high error rate were removed. With the one reading removed, the new $\bar{c}$ is 15.32; the new UCL_c is 27.06; and the new LCL_c is 3.58. Nevertheless, there are still points (for months 1 and 2) outside the upper control limit.

Even when you have recalculated the control limits, you see that there are still points outside the upper control limits for both B and C. Therefore, we can say the error rates for collectors B and C are not in control. Only the error rate for A is in control.

(b) Collector A has the lowest overall average count, $\bar{c}$. Collectors B and C have about the same overall average on the initial charts. When the $\bar{c}$ is recalculated, collector C has a better average than B.

(c) The charts do show that the error rates for new collectors decrease. This is especially true for B and C. It takes about five to seven months for the error rates to settle down.

(d) Six errors in two days is an upper specification for the error rates and was set by management. Not even the best collector, A, meets this specification. About half of the points still lie above this specification. Management's expectations appear to be unrealistic.

If the error rates for collectors B and C settle down and come into control, become stable, it appears the average rates ($\bar{c}$'s) would be around 20 for B and perhaps 12 for C. When c-charts are in control, responsibility for improvement rests with management, *not* the employees performing the task being charted. Therefore, if the error rates for the three toll collectors do become stable, responsibility for any improvement to better, lower rates rests squarely with management.

MODULE 7: SAMPLING PLANS

Problem 7-1.

This problem asks that you construct OC curves for Plan C ($n=150$, $c=0$) and Plan D ($n=150$, $c=1$) and evaluate the plans.

The OC curves and the calculations necessary to develop them are shown in Figure S-27.

The tables for each plan at the top of the figure show the development of the point values, which then determine the shape of the OC curve. The first column in each table is the percent (%) defective of a lot. The second column is that percent defective (p) value expressed as a decimal number. This is also called the fraction defective of the lot. The fraction defective (p) is multiplied by the sample size (n) to get the numbers in the column headed as np. With this number we will use Table 7-4 to obtain the probability of acceptance (P_a) values. Table 7-4 lists the probability of finding c (the acceptance number) or less defects in a sample that has an average number of defects equal to np.

The table for developing the OC curve for Plan C in Figure S-27 shows that for a lot percent defective of .25, the np value is .375. Referring to Table 7-4 we see that the first column is headed np. When we look down this column we find the closest we can come to .375 is .35 or .40. Our np value is halfway between these two values, so we will find the P_a by looking at the P_a values for .35 and .40 and choosing the value halfway between.

The acceptance number for Plan C is zero so the P_a number will be in the column headed by 0. It is the column immediately next to the np column. The P_a for .35 is .71 and the P_a for .40 is .67. Halfway between .67 and .71 is .69. This number has been recorded in the P_a column of the Plan C table.

Plan C shows the probability of acceptance (P_a) for .25 percent defective lots is .69. Plot this point by finding .25% on the horizontal scale, and then move straight up to .69 on the vertical scale. Continue finding P_a values in Table 7-4 for all the np values in the lot defective table of Figure S-27. The points represented by these P_a values are then marked on the OC curve portion of Figure S-27.

When you have placed enough points on the chart to determine the shape of the curve, you can draw the OC curve. Use the same procedure to draw the OC curve for Plan D. The points can be seen on the OC curves for Plan C and Plan D.

When both OC curves have been drawn, you can compare them to decide which is the better plan for meeting the goals of the department manager. These goals are (1) use a sampling plan that will accept lots of documents containing up to 25 errors per 10,000 documents (.25% defec-

Figure S-27. OC curves for Plan C and Plan D (Problem 7-1).

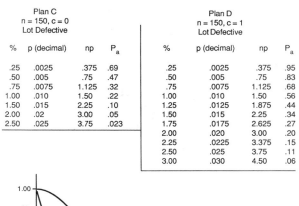

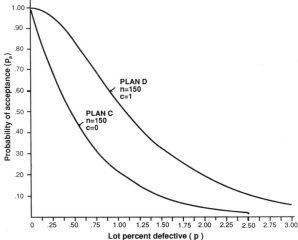

tive); and (2) reject lots containing 250 or more errors per 10,000 documents (2.5% defective).

The OC curve for Plan C shows that lots which are .25% defective have a probability of acceptance of .69. This means 69% of the lots will be accepted by the sampling plan and 31% will be rejected. The OC curve for Plan D shows that lots which are .25% defective have a probability of acceptance of .95. This means 95% of the lots will be accepted and 5% will be rejected.

In this comparison, Plan D is better than Plan C. Material that is .25% defective is to be considered acceptable and so should be accepted most of the time by the sampling plan used. Plan D accepts 95% of such lots and Plan C accepts 69%.

The second comparison to be made is the ability of the sampling plan to reject 2.5% defective material most of the time. The OC curve for Plan C shows that 2.5% defective lots have a probability of acceptance of .023, or 2.3% of the lots will be accepted and 97.7% will be rejected. The OC curve for Plan D shows a probability of acceptance of .10 for material that is 2.5% defective. This is the value traditionally used for the consumer's risk, that percent which is unwanted and should be rejected most of the time. The consumer's risk of 10% for Plan C falls at the 1.5% defective level instead of the 2.5% defective level as in Plan D. For this reason, Plan D is preferable.

The OC curves show that Plan C is a poorer plan because it will accept fewer acceptable lots (.25%) than will Plan D. Plan C is too strict because it rejects more than 90% of lots at the higher percent defective levels.

You should have selected Plan D as the better of the two sampling plans for meeting the goal of the department manager.

Problem 7-2.

This problem requires the construction of average outgoing quality (AOQ) curves for two different sampling plans. Plan C is $n=150$, $c=0$, and Plan D is $n=150$, $c=1$. The sampling plans are to be used with a lot size (k) of 1,000 items.

The calculation table and AOQ curve for Plan C are shown in Figure S-28. The calculations are made in decimals or fraction defective instead of percent values. Column 1 lists the fraction defective of various lots of material. Column 2 lists the probability of acceptance (P_a) of the various lot fraction defective values. The AOQ of the various lot fraction defective values is calculated by multiplying the value in Column 1 times the value in Column 2. The AOQ values are listed in fraction defective. The equivalent percent defective values are shown in parentheses.

The AOQ curve you constructed should look like the one shown in the lower portion of Figure S-28. The AOQ values in Column 3 are plotted to develop the AOQ curve. The AOQ for a lot fraction defective of .0025 is .0017. The point on the curve representing these two values is located by moving straight up from the .0025 fraction defective on the horizontal line of the chart to a point straight across from the value .0017 on the vertical scale, which is the AOQ fraction defective. The rest of the AOQ values in Column 3 are plotted on the chart. A curved line is drawn through these plotted points. This is the average outgoing quality curve.

The point on the AOQ curve with the largest AOQ value is the average outgoing quality limit (AOQL). Did you estimate the AOQL for Plan C to be approximately .0025? The calculation table and AOQ curve for Plan D are shown in Figure S-29. This curve was constructed using the same procedure as used for Plan C.

The curve you constructed for Plan D should look like the one in Figure S-29. The AOQL for Plan D can be estimated from this curve. Did you estimate the AOQL to be approximately .0056?

Figure S-28. AOQ curve for Plan C (Problem 7-2).

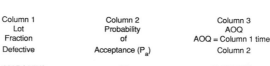

Plan C
n = 150, c = 0

Column 1 Lot Fraction Defective	Column 2 Probability of Acceptance (P_a)	Column 3 AOQ AOQ = Column 1 times Column 2
.0025 (.25%)	.69	.0017 (.17%)
.0050 (.50%)	.47	.00235 (.235%)
.0060 (.60%)	.41	.0246 (.246%)
.0075 (.75%)	.32	.0024 (.24%)
.0100 (1.0%)	.22	.0022 (.22%)
.0150 (1.5%)	.10	.0015 (.15%)
.0200 (2.0%)	.05	.0010 (.10%)

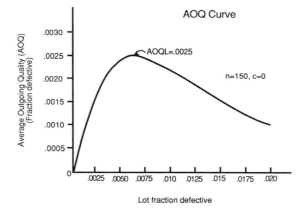

This problem asked if the AOQL estimates of the two sampling plans should change your answer to Problem 7-1. As the problem was stated, the department manager is concerned with the efficient acceptance of documents that are considered acceptable and the assured rejection of documents that should be corrected before leaving the department. No limit was placed on the quality level leaving the department other than the 250 errors per 10,000 documents, therefore the AOQL is not a primary consideration when evaluating the two sampling plans. Both plans show AOQLs less than 2.5%, which is 250 errors per 10,000 documents.

The information gained in this comparison of the AOQL of the two sampling plans is valuable information to have when things go wrong in the process. The AOQL represents the worst average outgoing quality, regardless of how bad things get in the process.

Figure S-29. AOQ curve for Plan D (Problem 7-2).

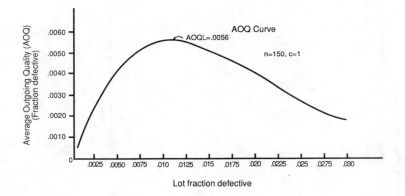

Plan D
n = 150, c = 1

Column 1 Lot Fraction Defective	Column 2 Probability of Acceptance (P_a)	Column 3 AOQ AOQ = Column 1 times Column 2
.0025 (.25%)	.95	.0024 (.24%)
.0050 (.50%)	.83	.00415 (.415%)
.0075 (.75%)	.68	.0051 (.51%)
.0100 (1.0%)	.56	.0056 (.56%)
.0125 (1.25%)	.44	.0055 (.55%)
.0150 (1.5%)	.34	.0051 (.51%)
.0175 (1.75%)	.27	.00473 (.473%)
.0200 (2.0%)	.20	.0040 (.40%)
.0225 (2.25%)	.15	.0034 (.34%)
.0250 (2.5%)	.11	.00275 (.275%)
.0300 (3.0%)	.06	.0018 (.18%)

Problem 7-3.

This problem asks that you compare the total inspection required by each of the two sampling plans described in Problem 7-1. The process average is known to be .25% defective and the lot size is 1,000 documents for both sampling plans.

You can compare the amount of documents that must be inspected to accept a given number of documents for shipment by looking at the results of submitting 100 lots of documents to each of the sampling plans.

Figure S-30 shows the sequence of events for each sampling plan when 100 lots are processed through the inspection system. The percent defective of the material is .25% and the 100 lots contain 100,000 documents. The documents are in lots of 1,000 documents.

Plan C is n = 150, c = 0. The probability of acceptance of .25% defective material with this plan is .69, therefore, we would expect 69 of the 100 lots to be accepted when sampled and inspected. The number of documents inspected in 69 lots is 69 times 150 which is 10,350 documents.

Figure S-30. Comparison of Plan C and Plan D (Problem 7-3).

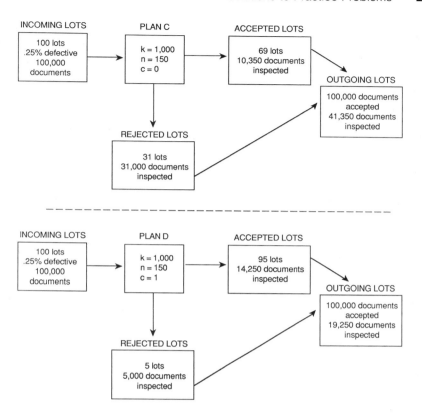

If 69 of the lots were accepted, then 31 lots were rejected by the sampling plan. All documents in these rejected lots must be inspected and corrected. This means 31,000 documents were inspected and made acceptable. When these documents were combined with the 69 lots that were accepted by inspecting 10,350 documents, we had 100,000 documents ready for shipment and had inspected a total of 41,350 documents.

Plan D is $n=150$, $c=1$. The probability of acceptance of .25% defective material is .95. We would expect 95 of the 100 lots to be accepted by the sampling plan. To accept the lots it was necessary to inspect 14,250 documents (95 times 150).

Five of the lots were rejected by the sampling plan. The 5,000 documents in these five lots were inspected, corrected, and merged with the accepted lots. We now had 100,000 documents ready for shipment. To accomplish this, we inspected a total of 19,250 documents.

If you followed the above procedure, you can easily answer which sampling plan requires the least total inspection. Plan C requires inspection of 41,350 documents out of 100,000. Plan D requires inspection of 19,250 documents out of 100,000

When considering the information obtained in Problems 7-1, 7-2, and 7-3 it becomes apparent that Plan D, n=150, c=1, is the most efficient plan, and it offers the needed protection against accepting unwanted material as well as the required protection against rejecting acceptable material.

Problem 7-4.

This problem asks you to recommend a sampling plan for John to use for acceptance sampling of towels received in the warehouse. The towels are received 500 to a box and 10,000 towels in a shipment. The supplier has agreed that .25% defective is an acceptable quality level.

Table 7-1 lists sample size code letters for use in selecting sampling plans from those listed in Table 7-2. These are plans of the type found in the Military Standard Sampling Procedures and Tables for Inspection by Attributes. These are widely used sampling plans that are available from the U.S. government. Before using these two tables, you must know the lot sizes of the material. In addition, you must select a level of inspection. The level chosen will have an effect on the size of the sample you will use.

Ten thousand towels are received in each shipment, but 500 in each box in the shipment. The question is, which should be chosen for the lot size—10,000 or 500? Looking at Table 7-1, we can see that a lot size of 500 using General inspection level II will result in sample size code letter H.

In Table 7-2, the first column contains the sample size code letters. Looking down this column you can find the letter H. The second column contains the sample size. If you use code letter H the sample size is 50 pieces. Moving horizontally across the table from the code letter H to the column headed 0.25, you will find the acceptance number for your sampling plan is zero. You now have a sampling plan that could be recommended to John for use in inspecting the towels. The plan is n=50, c=0, k=500.

This is not the only plan that you could select using the information at hand. Suppose you decide the lot size is 10,000. The towels are received 10,000 at a time. Look again at Table 7-1, but this time move down the "Lot or batch size" column to the 3,201 to 10,000 group. Move across to the General Inspection Level II column and you will find the code letter is L. referring to the sample size code letter column in Table 7-2, you will see that code letter L calls for a sample size of 200. When you move across to the column headed 0.25, you will find this time that the acceptance number is 1. So, now you have another sampling that John could use: n=200, c=1, k=10,000.

Both plans have an acceptable quality level of .25% defective. We can only assume that the incoming material will average .25% defective over the long term. The shipments of 10,000 towels will be accepted most of the time (about 90%) when either plan is used. One plan will have a sample size of 200 and will accept 10,000 towels of acceptable quality, while the other plan will require 20 samples of 50 each (1,000 towels) to accept 10,000 towels of the same quality. Since we have no other information about the process that produces the towels, the recommended plan should be n = 200, c = 1, k = 10,000.

When you make your recommendation to John, be sure to point out that when he selects his samples they should be collected in such a way that every towel in the 10,000-piece lot has an equal chance of being selected.

MODULE 8: SYSTEMS CAPABILITY
Problem 8-1.

This problem asks you to decide if the housekeeping system is capable of meeting a target value of 25 demerits maximum for cleaning the rooms in a hotel. The number is obtained by inspecting five rooms each day.

The first step in estimating the capability of this system is to develop an average and range chart to test for stability of the process and then, using the average range, calculate the upper limit for individual room values (UL_X). This value can then be compared with the target value to get an estimate of the capability.

In addition to these calculations, you should also compare the estimated total spread of the system with the tolerance or specification. The specification in this problem is 25 maximum, however, the lower specification limit can be considered to be zero.

Before completing the calculations examine the average and range chart for points out of control indicating the presence of assignable causes. The average and range chart you developed should look like the one in Figure S-31. There are no points out of control so the "LIMITS FOR INDIVIDUALS" section of the calculation worksheet on the back of the average and range chart can now be completed to determine the upper limit for individuals (UL_X) and the total spread of the process (6σ). See Figure S-32. These values are 31.96 for the upper limit for individuals and 28.64 for the total spread of the process.

Before accepting these values to answer whether the process is capable or not, it is well to confirm that the numbers used to develop the control chart are distributed about the average in a normal way. This is done by making a probability plot of the numbers. This plot is shown in Figure S-33. The fit of the plot points to a straight line, though not perfect, appears to be very good. You can, therefore, be confident of the estimates you will make using the calculations from the average and range chart.

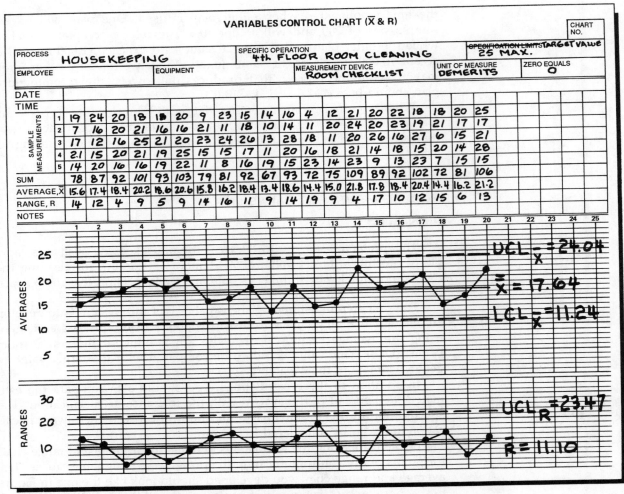

Figure S-31. Average and range chart for housekeeping system capability study (Problem 8-1).

At this point, you have enough information to say that the system as presently operating is not capable of meeting the target of 25 every time. No matter how hard housekeeping people try, the normal variation built into the system will result in some room demerit values that are greater than 25.

You were asked to determine the capability index for this system. The calculations you must make to arrive at the capability index are shown in Figure S-32. The capability index (C_{pk}) is the smaller of the two numbers obtained using the overall average minus the lower specification limit (LSL) divided by 3σ and the upper specification limit (USL) minus the overall average ($\bar{\bar{X}}$) divided by 3σ.

Figure S-32. Limits for individuals calculations, for housekeeping system capability study (Problem 8-1).

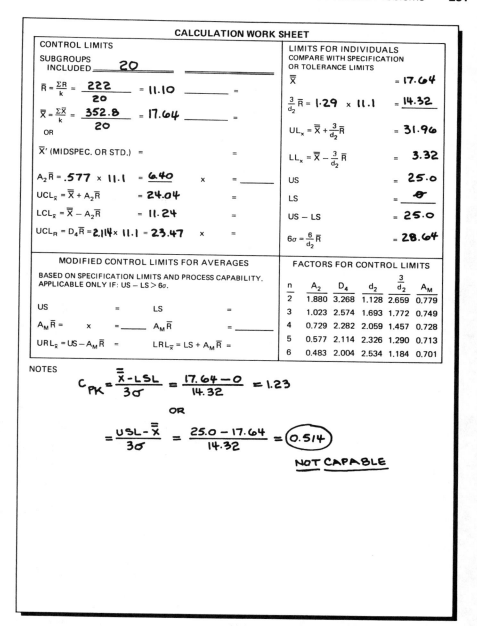

These two numbers are 1.23 and 0.514. The smaller of the two is 0.514, which is less than 1.00. The capability index must be 1.00 or greater before a process or system can be considered capable. Your answer should be that the capability index for this system is 0.514 and the system is NOT CAPABLE.

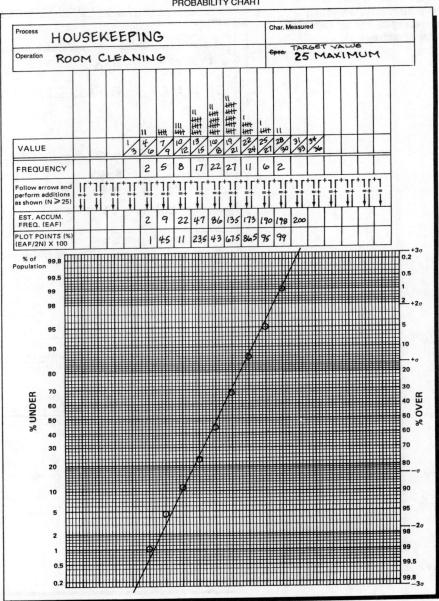

Figure S-33. Probability plot for housekeeping system capability study (Problem 8-1).

Glossary of Terms

A_2—A special factor used to calculate control limits for the averages on the average and range chart.

$\tilde{A}_2$—A special factor used to calculate control limits for the medians on a median and range chart. $\tilde{A}_2$ is not the same as A_2, which is used for average and range charts.

acceptable quality level (AQL)—The quality of material that will be accepted by a sampling plan most of the time (generally, 95% of the time). AQL is related to the producer's risk.

acceptance number (c)—The maximum number of defective items permitted in the sample that still allows the lot to be accepted for use without further inspection.

acceptance sampling—Selecting a number of items from a group or lot and using this sample to decide whether to accept or reject the entire lot.

AOQ—Average outgoing quality.

AOQL—Average outgoing quality limit.

AQL—Acceptable quality level.

assignable causes or special causes—Causes that the employee can do something about; detectable because they are not always active in the process. We can see assignable causes because they have a stronger effect on the process than chance causes.

attributes—Nonmeasurable characteristics. They are either present or they are not.

attributes chart—A type of chart in which characteristics are not measured in numbers but are considered acceptable or not acceptable, good or bad. The p-chart is an example of an attributes chart.

attributes or counting data—Data that come from nonmeasurable characteristics that can be counted.

average ($\overline{X}$)—The result of dividing the total or sum of a group of measurements by the number of things measured. Average is another term for "mean."

average and range chart—The most commonly used variables chart, also called $\overline{X}$-R (X bar, R) chart.

average outgoing quality (AOQ)—Long-term average quality (percent defective) of material present in lots after sampling inspection and screening.

average outgoing quality limit (AOQL)—The poorest overall average quality level obtained when rejected lots are screened and defective items are corrected or replaced with good items.

average range ($\overline{R}$)—The mean or the average value of a group of ranges on an average and range chart. It is used to calculate control limits for both averages and ranges on the average and range chart.

bell-shaped curve—See "normal distribution curve."

boundaries—A line between one interval and the next in the frequency histogram.

brainstorming—A group problem-solving method to bring out many ideas in a short time.

breakthrough—Deliberate action to achieve improvement.

c—(1) Symbol for the number of defects or errors in an inspection unit. (2) One of the basic parameters of any sampling plan. The acceptance number.

$\bar{c}$ (average number of defects for the process)—Equals the total of defects divided by the number of inspection units. $\bar{c}$ is the average number of defects for the process.

c-chart—A type of attributes control chart that helps monitor the number or count of errors or defects unit by unit, or by inspection units, in a service run. For example, we could count all errors in an order document.

capability index (C_p and C_{pk})—The number that expresses the capability of a process or system. (See "operation capability" and "system capability.") To find this index number, compare the process spread to the specification spread and express it in terms of the standard deviation. The C_p index does not take into account where the process is centered with respect to the required specifications.

cause and effect (C and E) diagram—A diagram that shows in picture or graph form how causes relate to the stated effect or to one another. Also known as a fishbone diagram.

chance causes or system causes—Causes that the employee can usually do nothing about because they are built into the process. Chance causes are continuously active in the process.

checksheet—A form for collecting data in an organized way so that the data are easy to analyze.

chronic problem——A type of problem that happens over and over again.

companywide quality control system—A system where all the people and elements of a company are involved with and concerned about the quality of the service being produced or delivered.

connector—A flow chart symbol that indicates a continuation of the process.

consumer's risk—The risk or probability of accepting bad material that is at the lot tolerance percent defective. It is customarily set at 10%.

continual improvement—An operating philosophy driven by top management that works to make every process in an organization better.

control—The prevention of change in a process. Also, the means used to maintain improved performance.

control chart—A special type of graph showing the results of periodic small inspections over time, like a movie of the process. A control chart tells when to correct or adjust a process and when to leave it alone.

control limits—Boundaries on a control chart within which the points plotted can vary without the need for correction or adjustment. Control limits are based on past performance and show what can be expected from a process as long as nothing changes.

corner count test—A simple test to determine whether or not a straight-line relationship exists between the two variables on a scatter diagram.

critical characteristics—Those characteristics of a service that enable the service to perform its designated function.

critical operations—Those operations that create critical characteristics.

criticality designation system—A system to determine which characteristics or elements of a service are most important to the customer. This system identifies those elements that require the use of statistical control techniques during the production and delivery of the service.

cumulative frequencies—Frequencies added successively on a Pareto diagram.

cumulative percentages—Cumulative frequencies on a Pareto diagram converted to percentages.

d_2—A special factor used with the average range ($\bar{R}$) to determine the standard deviation (σ).

D_4—A special factor used to calculate the upper control limits for ranges on an average and range chart.

$\tilde{D}_4$—A special factor used to calculate the upper control limits for the ranges on a median and range chart. $\tilde{D}_4$ is not the same as D_4, which is used for average and range charts.

decision—A flow-chart symbol indicating a place in the process being charted where an alternative action may be taken.

decision chart or decision flow chart—A type of chart that shows various steps in the decision process.

delay—A flow-chart symbol indicating that there is "waiting" before being able to go on to the next step in a service task or component.

dependent variable—The situation when one variable results from, or depends upon, another variable.

estimated accumulated frequency (EAF)—The accumulated frequency of measurements in a frequency distribution; used only to determine the plot points on a probability chart.

estimated process average—The value shown at the point where the line of best fit crosses the horizontal 50% line on the probability plot chart.

fishbone diagram—Also known as the cause and effect diagram.

floor solvable problem—A problem in which an assignable cause is present. Often a person on the job can find the assignable cause and correct it.

flow chart—See "process flow chart."

fraction defective p-chart—A p-chart that uses fractions instead of percentages. This p-chart shows how many units are defective compared to the total in the sample. The fractions are usually shown in decimal form.

frequency distribution—The pattern formed by a group of measurements of the same kind of units when the measurements are tallied according to how many times each one occurs.

frequency histogram or histogram—A "snapshot" of a process in block or bar graph form, showing the frequency of each measurement in a group of elements of a process for developing or delivering a service.

histogram—See "frequency histogram."

in control—A condition in which the points plotted on a control chart vary, but stay inside the control limits. When a process is in control, no assignable causes appear to be at work.

individual and range (X-R) chart—A type of variable control chart based on individual measurements rather than averages or medians of small samples. An X-R chart is helpful in monitoring situations where you are able to take only one reading during a time period and where measurements from the process follow the bell-shaped (normal) curve.

inherent variation—The natural variation in a process, due to chance causes.

in-process control system—A system that uses the practical techniques of statistical process control to measure, monitor, evaluate, and troubleshoot the processes that develop or deliver services.

inspection—A step on a process flow chart where someone checks or verifies that the service task or component meets the requirements.

inspection lot—A group or batch of items, material, or services from which a sample is taken. The lot will be accepted or rejected based on the quality of the sample.

inspection unit—May consist of one service unit, such as a document, or a group of units, such as a batch of checks from one account.

interval or class interval—A division on a frequency histogram marked off by boundaries, and all possible readings or measurements that can fall between those two boundaries.

k—One of the basic parameters of any sampling plan. The lot size.

line of best fit—A straight line drawn on a probability plot as close as possible to all the plot points.

lot size—The total number of units in the group from which we draw the sample.

lot tolerance percent defective (LTPD)—The limit of acceptable quality. The quality the customer or user wants rejected most of the time. LTPD is related to the consumer's risk.

lower control limit (LCL)—The lower boundary above which points plotted on a control chart can vary without the need for correction or adjustment.

lower limit for individuals (LL_X)—The estimated smallest individual value to be produced by the operation. This should not be confused with the lower control limit ($LCL_{\bar{X}}$) for averages.

lower specification limit (LSL)—The smallest acceptable value for the service unit or task produced by a process or operation.

LTPD—The lot tolerance percent defective.

management by detection—A traditional approach to quality control, which relies heavily on inspection and checking to maintain control of the process for producing and/or delivering a service. It removes errors but does not prevent them.

management solvable problem—A problem in which chance or system causes are at work. In order for this type of problem to be corrected, there must be basic changes in the process of delivering or producing the service. Management is primarily responsible for the solution.

mean ($\bar{X}$)—Another term for average.

median—The middle of a group of measurements counting from the smallest to the largest.

median and range ($\tilde{X}$-R) chart—A type of variables control chart that uses medians and ranges to determine whether a process needs to be corrected or should be left alone. It is best used in situations that follow the bell-shaped curve and that the employee can easily adjust.

median of medians ($\tilde{\tilde{X}}$)—The middle median of a group of medians, counting from the smallest to the largest; used to calculate control limits for medians.

median of ranges ($\tilde{R}$)—The middle range of a group of ranges on a median and range ($\tilde{X}$-R) control chart, counting from the smallest to the largest; used to calculate control limits for medians and ranges.

midpoint—The point of an interval that is an equal distance between the boundaries of an interval. The midpoint is found by dividing the width of the interval in half and adding this value to the lower boundary.

Mil-Std 105E—Sampling plans originally developed for use during World War II.

move—A step in the process flow chart where material, people, or information travel from one point in the process to another.

n—The sample size. In acceptance sampling, it is one of the basic parameters of any sampling plan.

normal distribution curve—A type of curve in which the measurements tend to cluster around the middle. Because this curve is shaped like a bell, it is sometimes called the bell-shaped curve.

normal probability paper—A special kind of graph paper used to record the probability plot.

n$\bar{p}$ (average number defective for the process)—The number of defective items in all samples divided by the number of samples taken.

np-chart—A type of attributes control chart that helps monitor the number of defective items in a service.

OC curve—Operating characteristic curve.

operating characteristic curve—The curve used to determine the probability of accepting a lot of material when the percent defective for the lot is known.

operation—A step in a process flow chart. The work required to complete a task.

operation capability—The short-term ability of an operation or part of a process to meet the specified quality characteristics. Usually measured by comparing the specifications to the spread (6σ) of the operation.

out of control—A condition in which the points plotted on a control chart go outside the control limits. This condition indicates that an assignable cause is at work, disrupting the process.

outgoing quality assurance system—A system that uses statistical tools to determine whether or not the process is running satisfactorily and is producing services that consistently meet the customers' specifications. This system acts as a check to make certain that all the other parts of the quality control system are working effectively.

overall mean ($\bar{\bar{X}}$)—The mean or average value of a group of averages from an average and range chart; used to calculate control limits for averages on the average and range chart.

$\bar{p}$—(1) average percent defective for the process. On a percent defective p-chart, it equals the total number of defectives divided by the total number of items inspected, then multiplied by 100%. $\bar{p}$ is expressed as a percentage. (2) average fraction or proportion defective for the process. On a fraction defective p-chart, it equals the total number of defectives divided by the total number of items inspected. $\bar{p}$ is expressed as a decimal fraction or proportion.

p-chart—A type of attributes control chart that helps to monitor or control the percent or fraction defective units in the operations that produce a service.

P_a—Probability of acceptance.

Pareto (pa-RAY-toe) analysis—Helps set priorities on which problems to solve first by sorting out the few really important problems from the more numerous but less important ones. Pareto analysis is useful for dealing with chronic problems.

Pareto diagram—A special type of bar graph that records the most frequent problem as the first bar, the next most frequent problem as the next bar, and so on. The Pareto diagram is the picture or graph part of Pareto analysis.

PDCA/PDCS—A basic model for continual improvement. PDCA stands for plan, do, check, and act. PDCS stands for plan, do, check, and standardize.

percent defective p-chart—A special type of attributes control chart. This p-chart shows the percentages of items that are defective or do not conform to specifications.

piggybacking—An activity in the brainstorming process by which people build on one another's ideas. It can be used as a prodding technique for brainstorming.

pinner—The person who posts the idea cards on the storyboard.

plot points—The points on the probability graph obtained from the frequency distribution of the measurements.

probability of acceptance (P_a)—The fraction or percentage of lots that will be accepted under a sampling plan for a given level of quality of material being sampled.

probability plot—A method for estimating how well the measurements used to make the average and range chart fit a normal curve. This method also tells what percent of the units may be outside the specifications.

process cause and effect diagram—A type of C and E diagram that follows the service through some or all of the steps to produce and/or deliver it.

process flow chart—A diagram that shows the steps of a particular job or task in sequence. It helps to track the flow of information, people, or paper through the system of delivering or producing a service.

process spread—The difference between the largest and smallest individual unit normally produced by an operation or process. When compared to the specifications, the process spread tells whether the process can produce or deliver services within the specifications. Also written as 6σ.

producer's risk—The probability or risk of rejecting good material that exactly meets the acceptable quality level. It is customarily set at 5%.

proportion defective p-chart—See "fraction defective p-chart."

R—The symbol for range.

$\bar{R}$—The symbol for average range.

random sample—A type of sample where each and every item in the group of items to be sampled has the same chance of being selected as part of that sample.

range—The difference between the smallest and the largest of a group of readings.

representative sample—a sample that has the best chance of reflecting the quality of the entire lot from which it is drawn.

sample—Several, but not all, of the possible readings in a group of one kind of item.

sample size (n)—In acceptance sampling, the number of items from a group that must be inspected in order to decide whether to accept or reject the whole group.

sampling bias—Any of a number of conditions when selecting a sample that will cause the sample not to be random, therefore, not representative of the material being sampled.

scatter diagram—A graph that shows how two variables may be related.

sigma (σ)—Symbol for standard deviation.

sporadic problem—A type of problem that happens only once in a while.

$\sqrt{}$ (square root)—Symbol for the square root, a special calculation that mathematicians have worked out.

stable process—A situation in which the variations in the service are due to chance causes alone. The service varies in a predictable manner.

standard deviation—A special calculation that describes how closely the measurements cluster around the middle of a normal curve. This number can be used to describe the process spread.

storage—A step in the process flow chart where work or material such as a printout is held before going to the next step.

storyboarding—A process for generating ideas where the ideas are written on cards and posted on a "board" under appropriate headings.

straight-line relationship—A situation where a change in one variable results in a change in the other variable and the amount of the change is always the same.

supplier control system—An effective way to control the quality of purchased materials; uses the criticality designation system.

system capability—The long-term ability of a process or system to produce or deliver a service according to the specified quality characteristics. It is concerned with the variation caused by all sources.

system causes—See "chance causes."

underlying frequency distribution—The pattern created by taking all possible items in a group of the same kind of items and arranging them in a frequency histogram. This pattern will always be the same because it includes all the items.

upper control limit (UCL)—The upper boundary below which points plotted on a control chart can vary without the need for change or correction.

upper limit for individuals (UL$_X$)—The estimated largest individual value to be produced by the operation. This should not be confused with the upper control limit (UCL$_{\overline{X}}$) for averages.

upper specification limit (USL)—The largest acceptable value for the service produced by a process or operation.

variable data—Data that come from things that can be measured. The measurements will vary from one element of a service to the next.

variables chart—A type of chart on which things or characteristics that are plotted are measured in numbers. The average and range ($\overline{X}$-R) chart is an example.

$\overline{X}$—The symbol for average on the average and range chart, pronounced "X bar."

$\overline{\overline{X}}$—The symbol for the average of averages of readings on the average and range chart. Also called the "overall mean."

$\overline{X}$-R chart—A type of variables control chart that uses averages and ranges to show whether the process needs to be adjusted or should be left alone.

Recommended Readings and Resources

1. Barra, Ralph. *Putting Quality Circles to Work: A Practical Strategy for Boosting Productivity and Profits.* New York: McGraw-Hill, 1983.

2. Charbonneau, Harvey C., and Webster, Gordon L. *Industrial Quality Control.* Englewood Cliffs, NJ: Prentice Hall, 1978.

3. Dodge, Harold F., and Romig, Harry G. *Sampling Inspection Tables: Single and Double Sampling.* New York: Wiley, 1959.

4. Feigenbaum, A. V. *Total Quality Control.* New York: McGraw-Hill, 1961.

5. Grant, Eugene L., and Leavenworth, Richard S. *Statistical Quality Control,* 6th ed. New York: McGraw-Hill, 1988.

6. Hayes, Glenn E., and Romig, Harry G. *Modern Quality Control.* Encino, CA: Glencoe, 1982.

7. Imai, Masaaki. *Kaizen (Ky'zen): The Key to Japan's Competitive Success.* New York: Random House, 1986.

8. Ishikawa, Kaoru. *Guide to Quality Control.* White Plains, NY: Quality Resources, 1984.

9. Ishikawa, Kaoru. *What is Total Quality Control? The Japanese Way.* David J. Lu, translator. Englewood Cliffs, NJ: Prentice Hall, 1985.

10. Juran, J. M., and Gryna, Frank M. *Quality Planning and Analysis: From Product Development Through Use.* New York: McGraw-Hill, 1985.

11. McNellis, Jerry. *An Experience in Creative Thinking.* New Brighton, PA: The Creative Planning Center. For more information on storyboarding, contact the Center at 519 Ninth Street, New Brighton, PA 15066; 412-847-2120.

12. Military Standard 105E. *Sampling Procedures and Tables for Inspection by Sampling.* Washington, D.C.: U.S. Government Printing Office, 1989.

13. Olmstead, Paul S., and Tukey, John W. "A Corner Test for Association," *Annals of Mathematical Statistics,* Vol. 18, Dec. 1947, pp. 495–513.

14. Ott, Ellis R. *Process Quality Control: Troubleshooting and Interpretation of Data.* New York: McGraw-Hill, 1975.

15. Shewhart, Walter A. *Economic Control of Quality of Manufactured Product.* New York: Van Nostrand, 1931. Republished in 1980 by the American Society for Quality Control. This is the original book on quality control.

16. Walton, Mary. *The Deming Management Method.* New York: Dodd, Mead, 1986.

Appendix: Factors and Formulas

This appendix contains the factors and formulas for calculating control limits for all the control charts described in this book.

Factors and formulas for average and range charts.

Sample size, n	A_2	D_4
2	1.880	3.268
3	1.023	2.574
4	0.729	2.282
5	0.577	2.114
6	0.483	2.004

Upper control limit for averages ($\overline{X}$):
$$UCL_{\overline{X}} = \overline{\overline{X}} + (A_2 \text{ times } \overline{R})$$
Lower control limit for averages ($\overline{X}$):
$$LCL_{\overline{X}} = \overline{\overline{X}} - (A_2 \text{ times } \overline{R})$$
Upper control limit for ranges (R):
$$UCL_R = D_4 \text{ times } \overline{R}$$
Lower control limit for ranges is zero for samples of six or less.

Factors and formulas for median and range charts.

Sample size, n	$\widetilde{A}_2$	$\widetilde{D}_4$
2	2.22	3.87
3	1.26	2.75
4	0.83	2.38
5	0.71	2.18

Upper control limit for medians ($\widetilde{X}$):
$$UCL_{\widetilde{X}} = \widetilde{\overline{X}} + (\widetilde{A}_2 \text{ times } \widetilde{R})$$
Lower control limit for medians ($\widetilde{X}$):
$$LCL_{\widetilde{X}} = \widetilde{\overline{X}} - (\widetilde{A}_2 \text{ times } \widetilde{R})$$
Upper control limit for ranges (R):
$$UCL_R = \widetilde{D}_4 \text{ times } \widetilde{R}$$
Lower control limit for ranges is zero for samples of six or less.

Factors and formulas for individual and range charts

Sample size, n	Factor for individuals	D_4
1	2.66	3.268

Upper control limit for individuals (X):
$$UCL_X = \overline{\overline{X}} + (2.66 \text{ times } \overline{R})$$
Lower control limit for individuals (X):
$$LCL_X = \overline{\overline{X}} - (2.66 \text{ times } \overline{R})$$
Upper control limit for ranges (R):
$$UCL_R = D_4 \text{ times } \overline{R}$$
Lower control limit for ranges is zero.

Formulas for p-charts

Upper control limit for p (percent defective):
$$UCL_p = \overline{p} + 3\sqrt{\frac{\overline{p} \times (100\% - \overline{p})}{n}}$$
Lower control limit for p (percent defective):
$$LCL_p = \overline{p} - 3\sqrt{\frac{\overline{p} \times (100\% - \overline{p})}{n}}$$

Upper control limit for p (fraction defective):

$$UCL_p = \bar{p} + 3\sqrt{\frac{\bar{p} \times (1-\bar{p})}{n}}$$

Lower control limit for p (fraction defective):

$$LCL_p = \bar{p} - 3\sqrt{\frac{\bar{p} \times (1-\bar{p})}{n}}$$

Formulas for np-charts

Upper control limit for np (number defective):

$$UCL_{np} = n\bar{p} + 3\sqrt{n\bar{p} \times (1 - n\bar{p}/n)}$$

Lower control limit for np (number defective):

$$LCL_{np} = n\bar{p} - 3\sqrt{n\bar{p} \times (1 - n\bar{p}/n)}$$

Formulas for c-charts

Upper control limit for c (count of defects):
$$UCL_c = \bar{c} + 3\sqrt{\bar{c}}$$

Lower control limit for c (count of defects):
$$LCL_c = \bar{c} - 3\sqrt{\bar{c}}$$

Factors and formulas for making capability estimates

Sample size, n	d_2
2	1.128
3	1.693
4	2.059
5	2.326
6	2.534

Upper limit for individuals:
$$UL_X = \bar{\bar{X}} + 3\bar{R}/d_2$$

Lower limit for individuals:
$$LL_X = \bar{\bar{X}} - 3\bar{R}/d_2$$

Formula for process spread (operation capability):
$$6\sigma = 6\bar{R}/d_2$$

Formula for capability index:

$$C_p = \text{tolerance}/6\sigma$$
$$C_{pk} = \text{smaller of } (USL - \bar{\bar{X}})/3\sigma \text{ or } (\bar{\bar{X}} - LSL)/3\sigma$$

Index

A_2 factor, 117-18
$\tilde{A}_2$ factor, 126, 127, 128
Acceptable quality level (AQL), 190-91, 199-200, 204
Acceptance number (c), 179-80, 200, 201-2, 204, 205
   changing, 191-93
Acceptance sampling, 174-77, 203, 204
Alpha (α). *See* Producer's risk
AOQ. *See* Average outgoing quality
AOQ curve. *See* Average outgoing quality curve
AOQL. *See* Average outgoing quality limit
AQL. *See* Acceptable quality level
Assignable causes, 6-7, 8, 12-14, 18, 20, 21, 94, 103, 116, 119, 135, 144, 165, 207, 211, 227
   sources of, 106, 145-46, 166
Attribute or counting data, 137
Attribute plans, 178-79
   see also Sampling plans
Attributes control charts, 10, 12-14, 137-38, 171
   c-charts, 162-70
   np-charts, 159-62
   p-charts, 138-59
Average ($\overline{X}$), 8, 10, 100, 101
   outside control limits, 104-5, 116-20
Average, overall. *See* Overall average
Average and range ($\overline{X}$-R) charts, 10-11, 12, 99-122, 135, 209, 210-11, 216, 227-28, 231, 232
   control limits for, 113-15, 116-18
   interpreting, 103-7
   setting up, 107-20

Average and range charts (*continued*)
   tests for in-control, 115-16, 118-20
   using, 101-2
      in continued operations, 121-22
Average number of defects. *See* $\overline{c}$
Average outgoing quality (AOQ), 194-95, 205
Average outgoing quality (AOQ) curve, 182, 194-203
   developing, 195-98
Average outgoing quality limit (AOQL), 198, 203, 205
Average range ($\overline{R}$), 112-13, 116, 211, 214

Bar graph, 40
   see also Pareto diagram
Bell-shaped curve. *See* Normal distribution curve
Beta (β). *See* Consumer's risk
Boundaries, 78-79
Brainstorming, 21, 25-38, 47, 59, 234, 235, 237
   and cause and effect diagrams, 32-38
   handling difficulties with, 32
   preliminary steps for, 26-27
   procedure for, 27-29
   prodding techniques for, 29-31
Breakthrough, xvii-xviii

c (acceptance number). *See* Acceptance number
c (number of defects), 162
$\overline{c}$ (average number of defects), 168
C and E diagrams. *See* Cause and effect diagrams

Capability, 121, 208-10
    see also Operation capability; System capability
Capability index (C_p and C_{pk}), 228-31, 232
Capability studies, 207-32, 235, 236
    and capability indexes, 228-31
    operation (short-term), 210-26
    system, 227-28
Cause and effect diagrams, 21, 32-38, 47-48, 235, 236-37
    constructing, 34-38
    types of, 38
    uses of, 34
    see also Fishbone diagrams
Causes of variation. See Assignable causes; Chance causes; Inherent causes; System causes
c-charts, 138, 162-70, 171
    control limits for, 168-70
    interpreting, 164-66, 170
    setting up, 166-70
    tests for in-control, 170
    using, 163-64
Chance causes, 6-7, 8, 12-14, 165
Charts. See names of specific charts
Checklist checksheet, 91-92, 94
Checksheets, 89-93, 94, 235
    checklist, 91-92
    frequency histogram, 90-91
    item, 92
    location, 92-93
    matrix, 93
Chronic problems, 25, 31, 38
Class intervals. See Intervals
Companywide quality control system, 15
Connector symbol, 53
Consumer's risk (β), 177, 180, 183, 184-85, 199, 204, 205
Continual improvement, xv-xvii, 94, 194, 204
Control, xvii
    see also Control charts; Control limits; In control; Out of control; Statistical control; Statistical process control

Control charts, 9-14, 17, 99, 134-35, 235, 236
    attributes, 12-14, 137-72
        c-charts, 138, 162-70
        np-charts, 138, 159-62, 171
        p-charts, 138-59, 171, 204
    management review of, 21-22
    purposes of, 107
    uses of, 2-3, 21
    variables, 10-12, 99-136
        average and range, 10-11, 99-122
        individual and range, 129-34
        median and range, 122-29
Control limits, 9-10, 12, 100, 102, 121, 122, 135, 171, 207, 211, 212
    calculations for
        average and range chart, 113-15, 116-18
        c-chart, 168-70
        individual and range chart, 131-32
        median and range chart, 126-28
        np-chart, 160-62
        p-chart, 152-53, 156
    points outside, 104-5, 106-7, 128, 133-34, 144, 157-59, 164-66, 170, 171
Corner count test, 65-69
"Corrective Action Process," 233
Counting data. See Attribute or counting data
C_p. See Capability index
C_{pk}. See Capability index
Critical characteristics, 16
Criticality designation system, 15-17, 18, 108
Critical operations, 16-17
Cumulative frequency, 45
Cumulative percentage, 45
Curves
    average outgoing quality, 182, 194-203
    normal distribution, 4-5, 6, 8, 122, 135, 232
    operating characteristic, 182-94
Customers, xiv, xv, xvi, 1-2, 17, 22
    internal, xv

d_2 factor, 214
D_4 factor, 114-15
$\tilde{D}_4$ factor, 126, 127
Data, variable, 63, 64
Decision charts
   for working with averages, 120
   for working with p's, 157-58
   for working with ranges, 116-17
Decision making, 19-20
Decision symbol, 53
Defects. *See* c; $\bar{c}$
Delay symbol, 52
Dependent variable, 64
Diagrams
   fishbone, 6
   Pareto, 39-47, 235, 236, 237
   scatter, 62-69, 70, 235
   see also Cause and effect diagrams

EAF. *See* Estimated accumulated frequency
Employees, 6-7, 20, 31, 106, 146
   involvement in quality control, 14
   problem solving and, 21
Environment, 6-7, 31, 106, 146
Equipment, 6-7, 31, 106, 145-46
Estimated accumulated frequency (EAF), 220-21
Estimated process average, 223-25

Factors
   for average and range charts, 113-15, 116-18
   for median and range charts, 126-27
   for individual and range charts, 132
   for standard deviation, 213-14
Fishbone diagrams, 6, 38
   see also Cause and effect diagrams
Floor solvable problem, 20, 21
Florida Power & Light, 233
Flow charts, 21, 51-56, 116
   see also Decision charts; Process flow charts

Fraction defective p-charts, 153-57
   control limits for, 156
   see also Percent defective p-charts
Framework for solving problems, 233-37
Frequency, cumulative, 45
Frequency distribution, 4, 5, 8, 9, 85-87, 135, 216, 217, 219
   histogram as a picture of, 74-75, 87
   tallying for histogram, 79-80, 81
   underlying, 85-87, 88-89
   and variation, 6-7, 73-74, 85-87
Frequency histogram, 74-89, 94, 211, 236
   checksheet, 89-91
   constructing, 75-82
      cautions, 82-85
   interpreting, 87-91
   relation to underlying frequency distribution, 85-87
   uses of, 8-9, 19, 87-91
   and variation, 74-75, 76

Guide to Quality Control (Ishikawa), 105

Histogram, 8
   see also Frequency histogram

Improvement. *See* Continual improvement
In control, 12, 18, 164
Individual and range (X-R) charts, 129-34, 135
   control limits for, 131-32
   developing, 130-31
   tests for in-control, 133-34
   using, 129-30
      in continued operations, 134
Inherent causes. *See* Chance causes
Inherent variation, 103, 106, 107, 108, 109, 113, 116, 121, 122, 135, 165
In-process control system, 17-18, 22
Inspection lot, 181-82
Inspection symbol, 52

Inspection unit, 162
Internal customers, xv
Intervals, 78-79, 82-84
Item checksheet, 92, 94

Japan, xiii, 233

k. *See* Lot size

LCL. *See* Lower control limit
Limits for individuals, 212-17
Line of best fit, 223, 225-27
LL$_X$. *See* Lower limit for individuals
Location checksheet, 92-93, 94
Long-term capability. *See* System capability
Lot size (k), 179-80
Lot tolerance percent defective (LTPD), 190-91, 198-200, 204
Lower control limit (LCL), 11, 102, 115, 116
  average and range chart, 115, 118-19, 122
  c-chart, 170
  individual and range chart, 131-34
  median and range chart, 126-28
  np-chart, 160-62
  p-chart, 152-53, 156, 158-59
Lower limit for individuals (LL$_X$), 212-17, 225-27, 231
Lower specification limit (LSL), 214, 230, 231
LSL. *See* Lower specification limit
LTPD. *See* Lot tolerance percent defective

Management, state-of-the-art, 14-15, 18, 21
Management by detection, 18
Management solvable problem, 20, 21
Managers, 20, 21, 22
Materials, 6-7, 31, 106, 146

Matrix checksheet, 93, 94
Mean. *See* Average
Median
  for median and range chart ($\widetilde{X}$), 124
  for scatter diagram, 66-67, 68
Median and range ($\widetilde{X}$-R) charts, 122-29, 135
  control limits for, 126-28
  developing, 123-28
  tests for in-control, 126, 128
  using, 122
    in continued operations, 128-29
Median of medians ($\widetilde{\widetilde{X}}$), 125
Median of ranges ($\widetilde{R}$), 125
Midpoints, 78-79, 87
Military Standard (Mil-Std) 105E Standard Sampling Tables, 175-76, 179
Move symbol, 52

n. *See* Sample size
Normal curve. *See* Normal distribution curve
Normal distribution curve, 4-5, 6, 8, 122, 135, 232
Normal probability paper, 219
n$\bar{p}$ (average number defective for the process), calculating, 160
np-chart, 138, 159-62, 171
  control limits for, 160-62
  interpreting, 162
  setting up, 159-62

OC curve. *See* Operating characteristic curve
Operating characteristic (OC) curve, 182-94, 199-201, 204
  developing, 185-90
  interpreting, 184-85
Operation capability, 208, 210-27, 231-32
  and average and range charts, 210-12
  and capability indexes, 228-31
  and limits for individuals, 212-17
  and probability plot, 217-27

Operation symbol, 52
Outgoing quality assurance system, 18-19
Out of control, 11, 105, 119, 145
Overall average ($\bar{\bar{X}}$), 111, 116, 119, 121, 122, 207, 214, 215
Overall mean. See Overall average

p (fraction defective), calculating, 153-55
p (percent defective), 190-92, 205
   calculating, 138, 141, 149
$\bar{p}$ (average fraction defective), 156
$\bar{p}$ (average percent defective), 145, 149, 151, 152, 158, 159
P_a. See Probability of acceptance
Pareto analysis, 21, 38-47, 48, 234
   purpose of, 38-40, 46
   relation to Pareto diagram, 40, 48
Pareto diagram, 39-47, 235, 236, 237
   constructing, 40-46
   interpreting, 46-47
   relation to Pareto analysis, 40, 48
p-charts, 138-59, 171, 204
   control limits for, 152-53, 156
   fraction defective, 153-57
   interpreting, 143-45, 157
   percent defective, 138-53
   setting up, 146-57
   tests for in-control, 157
   using, 138-43
      in continued operations, 157-59
PDCA/PDCS, xvi-xvii
Percentage, cumulative, 45
Percent defective (p). See p (percent defective)
Percent defective p-charts, 12, 138-53
   control limits for, 143-46, 152-53
   interpreting, 143-46
   setting up, 146-53
   using, 138-43
Piggybacking, 29
Pinner, 57, 58
Plot points, 221-23
Practice problems. See Problems

Probability of acceptance (P_a), 180, 184-85
   determining, 186-89
Probability plot, 19, 209, 217-27, 228, 231-32
Problems
   attributes control charts, 171-72, 266-71
   basic principles, 23
   frequency histograms and checksheets, 94-97, 249-58
   quality improvement tools, 70-71, 243-48
   quality problem-solving tools, 48-49, 239-43
   sampling plans, 205-6, 272-79
   solutions, 239-82
   systems capability, 232, 279-82
   variables control charts, 136, 259-66
Problem solving, 19-20
   see also Framework for solving problems
Problem-solving tools, 21, 25-49
Process
   in control, 12, 18, 164
   out of control, 11, 105, 119, 145
Process capability. See Capability
Process capability studies. See Capability studies
Process cause and effect diagrams, 38, 48, 235
Process flow charts, 51-56, 69, 234, 235
   constructing, 53-55
   symbols for, 52-53
   using, 55-56
Process Quality Control (Ott), 105
Process spread, 8, 210, 227, 228, 231
   estimating, 223-25
Producer's risk (α), 177, 180, 183, 184, 199, 204, 205
Proportion defective p-chart. See Fraction defective p-chart

"QC Story," 233
"QI Story," 233

Quality
  achieving, xv-xvii
  control system, 14-15
  definition, xiv-xv
  improvement tools, 51-71
  tools of, 7-22; *see also* Control charts; Frequency histogram; Probability plot

R. *See* Range
$\bar{R}$. *See* Average range
$\tilde{R}$. *See* Median of ranges
Random sampling, 181-82
Range, 10, 100, 135, 228
  calculating average, 112-13
  outside control limits, 106-7, 115-16, 126, 133-34
Range charts. *See* Average and range charts
Representative sampling, 181
Risks, sampling, 177, 204

Sample size (n), 179-80, 200, 201-2, 204, 205
  changing, 193-94
Sampling bias, 181
Sampling plans, 173, 177-203
  applications of, 177-78
  limitations of, 180
  parameters for, 179-80
  risks of, 177, 204
  selecting, 198-203
  standard, 179
Sampling techniques, 181-82
Scatter diagrams, 62-69, 70, 235
  constructing, 63-65
  corner count test, 65-69
Service industry, xiii-xiv, xvii, 1-2, 174
Short-term capability. *See* Operation capability
Sigma (σ), 8, 214, 217, 231
  see also Standard deviation

Solutions, practice problems, 239-82
  attributes control charts, 266-71
  frequency histograms and checksheets, 249-58
  quality improvement tools, 243-48
  quality problem-solving tools, 239-43
  sampling plans, 272-79
  systems capability, 279-82
  variables control charts, 259-66
SPC. *See* Statistical process control
Specifications
  average and range, 121
  capability index, 228-31
  capability studies, 210-17
  control limits versus, 121, 135
  frequency histograms, 87-89
  probability plot, 225-27
  see also Lower specification limit; Upper specification limit
Sporadic problems, 25
Spread. *See* Process spread
SQC. *See* Statistical quality control
Square root ($\sqrt{}$), 152-53
Stable process, 7, 8, 207, 211, 214-15, 216, 228
Standard deviation, 8, 210, 231, 232
  see also Sigma (σ)
Standard sampling plans, 179
State-of-the-art management and managers, 14-15, 18, 21
Statistical control, 2, 12-13, 102, 103-5, 115-16, 118-20, 121, 135, 142, 157-58
Statistical process control (SPC)
  basic principles of, 2-5
  purposes of, 20
  tools and techniques, 7-22, 25-49, 51-71
  see also Control charts; Frequency histogram; Sampling plans
Statistical Quality Control (Grant and Leavenworth), 105
Storage symbol, 52
Storyboarding, 21, 56-62, 69
  evaluating, 61-62
  preliminary steps for, 57-58

Storyboarding (*continued*)
  procedures for, 59-61
  uses of, 56-57, 235, 236-37
Straight-line relationship, 65
Supplier control system, 17
Symbols, flow chart, 52-53
System capability, 207, 208, 210, 227-32
System causes. *See* Chance causes

Tables, sampling, 175-76, 179
Tools of quality, 7-22
  framework for solving problems, 234-37
  problem-solving, 25-49
  quality improvement, 51-71
  see also Control charts; Frequency histogram; Probability plot

UCL. *See* Upper control limit
UL_X. *See* Upper limit for individuals
Underlying frequency distribution, 86, 87, 94
Upper control limit (UCL), 11, 12, 102, 114, 115, 116
  average and range chart, 114-15, 117-18, 122
  c-chart, 168
  individual and range chart, 131-134
  median and range chart, 126-28
  np-chart, 160-62
  p-chart, 152-53, 156, 158-59

Upper limit for individuals (UL_X), 212-17, 225-27, 231
Upper specification limit (USL), 214, 230-31
USL. *See* Upper specification limit

Variability, 19
Variable data, 63, 64, 137
Variables control charts, 10-12, 99-135
  average and range ($\bar{X}$-R) charts, 99-122
  individual and range (X-R) charts, 129-34
  median and range ($\tilde{X}$-R) charts, 122-29
Variation, 74-75, 87, 90, 108, 204, 227
  definition, 3, 73-74, 94
  patterns of, 4-5, 94
  sources of, 6-7
  see also Assignable causes; Chance causes; Inherent variation

$\bar{X}$. *See* Average
$\bar{\bar{X}}$. *See* Overall average
$\tilde{X}$. *See* Median
$\tilde{\tilde{X}}$. *See* Median of medians
X-R chart. *See* Individual and range charts
$\bar{X}$-R chart. *See* Average and range charts
$\tilde{X}$-R chart. *See* Median and range charts